# Studies in Fuzziness and Soft Computing

## Volume 340

**Series editor**

Janusz Kacprzyk, Polish Academy of Sciences, Warsaw, Poland
e-mail: kacprzyk@ibspan.waw.pl

*About this Series*

The series "Studies in Fuzziness and Soft Computing" contains publications on various topics in the area of soft computing, which include fuzzy sets, rough sets, neural networks, evolutionary computation, probabilistic and evidential reasoning, multi-valued logic, and related fields. The publications within "Studies in Fuzziness and Soft Computing" are primarily monographs and edited volumes. They cover significant recent developments in the field, both of a foundational and applicable character. An important feature of the series is its short publication time and world-wide distribution. This permits a rapid and broad dissemination of research results.

More information about this series at http://www.springer.com/series/2941

Jagdeep Kaur · Amit Kumar

# An Introduction to Fuzzy Linear Programming Problems

Theory, Methods and Applications

Springer

Jagdeep Kaur
School of Mathematics Punjab
Thapar University
Patiala
India

Amit Kumar
School of Mathematics Punjab
Thapar University
Patiala
India

ISSN 1434-9922       ISSN 1860-0808   (electronic)
Studies in Fuzziness and Soft Computing
ISBN 978-3-319-81003-4      ISBN 978-3-319-31274-3   (eBook)
DOI 10.1007/978-3-319-31274-3

Printed on acid-free paper

This Springer imprint is published by Springer Nature
The registered company is Springer International Publishing AG Switzerland

*Dedicated to the Almighty
and to My Parents*

# Preface

Decision making is the process of identifying and choosing alternatives based on the values and preferences of the decision maker. Decision making is also the process of sufficiently reducing uncertainty and doubts about alternatives to allow a reasonable choice to be made from among them. Optimization is a kind of decision making in which decisions have to be taken to optimize one or more objectives under some prescribed set of circumstances. The optimization model under the consideration of linear objective function and linear constraints becomes a linear programming model.

Over the past several decades linear programming models have been primarily developed in a deterministic, crisp environment. In such models the objectives and constraints are formulated in a 'hard' crisp manner leaving no scope for uncertainty and vagueness. Any linear programming model representing real-world situations involves a lot of parameters whose values are assigned by experts. However, both experts and decision makers frequently do not precisely know the value of these parameters. In such situations, it is practical to develop linear programming models of the realistic problem in a fuzzy environment and the linear programming problem under consideration becomes a fuzzy linear programming problem.

In the past few years, different methods have been proposed to find the fuzzy optimal solution of fully fuzzy linear programming problems in which all the parameters and decision variables are represented by nonnegative fuzzy numbers.

However, very few methods have been proposed to find the fuzzy optimal solution of such fully fuzzy linear programming problems in which all the parameters and/or decision variables are represented by unrestricted fuzzy numbers. The aim of this book is to present the existing methods for solving such fully fuzzy linear programming problems at one place in which some or all the parameters and/or variables are represented by unrestricted fuzzy numbers. This book presents the preliminaries, fundamentals, and methods to find the optimal solution, optimality of fully fuzzy linear programming problems, and the comparative study.

The contents of the book are divided into seven chapters. In Chap. 1, state of the art and origin of the fully fuzzy linear programming problems are presented. In

Chap. 2, the method to find the fuzzy optimal solution of such fully fuzzy linear programming problems with equality constraints in which all the variables are represented by nonnegative trapezoidal fuzzy numbers while all the other parameters are represented by unrestricted trapezoidal fuzzy numbers are presented. In Chap. 3, the method to find the fuzzy optimal solution of such fully fuzzy linear programming problems with equality constraints in which all the parameters and variables are represented by unrestricted trapezoidal fuzzy numbers is presented. In Chap. 4, the method to find the fuzzy optimal solution of such fully fuzzy linear programming problems with equality constraints in which all the parameters and variables are represented by unrestricted *LR* flat fuzzy numbers is presented. In Chap. 5, the method to find the fuzzy optimal solution of such fully fuzzy linear programming problems with inequality constraints in which all the parameters and variables are represented by unrestricted *LR* flat fuzzy numbers is presented. In Chap. 6, the method to find the unique fuzzy optimal solution of such fully fuzzy linear programming problems with equality constraints in which all the parameters are represented by unrestricted *LR* flat fuzzy numbers is presented. In Chap. 7, based on the present study the future work has been suggested.

The authors thank the Series Editor Janusz Kacprzyk for the insightful comments and suggestions. The first author acknowledges the financial support given by the UGC Dr. D.S. Kothari Post Doctoral Fellowship Scheme (No. F.4-2/2006 (BSR)/13-1134/2013(BSR)). The second author acknowledges the adolescent blessings of Mehar (lovely daughter of his cousin Dr. Parmpreet Kaur). He believes that Mata Vaishno Devi has appeared on the earth in the form of Mehar and without her blessings it would not be possible to think of the ideas presented in this monograph.

# Contents

# Abstract

Decision making is the process of making a choice between a number of options and committing to a future course of actions. Making a decision implies that there are alternative choices to be considered, and in such a case it is desired not only to identify as many of these alternatives as possible but to choose the one that (1) has the highest probability of success or effectiveness and (2) best fits with our goals, desires, lifestyle, values, and so on. Decision making is also the process of sufficiently reducing uncertainty and doubts about alternatives to allow a reasonable choice to be made from among them.

Optimization is a kind of decision making in which decisions have to be taken to optimize one or more objectives under some prescribed set of circumstances. The problems that often arise as a result of mathematical modeling of many real-life situations are called optimization problems. Optimization problems are frequently encountered in disciplines, such as water resources management, finance, engineering optimization, manufacturing systems, management, business, physical sciences, agriculture, etc.

The most general form of optimization problems is:

$$\text{Minimize/Maximize } f^\rho(X), X = (x_1, x_2, \ldots, x_n)$$

subject to

$$g_j(X) \leq 0 \, \forall j = 1, 2, \ldots, k$$

$$h_j(X) = 0 \, \forall j = 1, 2, \ldots, m$$

where $f, g_1, g_2, \ldots, g_k, h_1, h_2, \ldots, h_m$ are real valued functions defined on $\mathbb{R}^n$ and $n$ is a positive integer. The function $f$ is usually called the objective function. Each of the constraints $g_j(X) \leq 0$ is called an inequality constraint, and each of the constraints $h_j(X) = 0$ is called an equality constraint. Constraints of the type $g_j(X) \geq 0$ may be written as $-g_j(X) \leq 0$ and therefore they are not mentioned separately. $X = (x_1, x_2, \ldots, x_n) \in \mathbb{R}^n$ is called a decision vector and $x_1, x_2, \ldots, x_n$

are called decision variables or unknown variables. If $\rho = 1$ and all the functions $f^\rho(X), g_j(X)$ and $h_j(X)$ are linear then the problem is called a single-objective linear programming problem.

Over the past several decades optimization models have been primarily developed in a deterministic, crisp environment. In such models the objectives and constraints are formulated in a 'hard' crisp manner leaving no scope for uncertainty and vagueness. However, in real-life situations there may exist uncertainty. In such situations it is practical to develop mathematical models of the realistic problem in a fuzzy environment and the optimization problem under consideration becomes a fuzzy programming problem.

Any linear programming model representing real-world situations involves a lot of parameters whose values are assigned by experts. However, both experts and decision makers frequently do not precisely know the value of these parameters. Therefore, it is useful to consider the knowledge of experts about the parameters as fuzzy data (Zadeh, L.A.: Information and Control 8, 338–353, 1965).

This book is an academic monograph based on the papers published in popular international journals by the authors. The focus of this book is to present the theory, methods, and the applications of fully fuzzy linear programming problems and present all the methods for solving fully fuzzy linear programming problems at one place. The book includes seven chapters.

Chapter 1 presents a brief review of the work done in the area of linear programming problems with fuzzy parameters.

In Chap. 2, it is depicted that Lotfi et al. (Applied Mathematical Modelling 33, 3151–3156, 2009) pointed out that there is no method in the literature for solving fully fuzzy linear programming problems with equality constraints and a method is presented for the same. The limitations and shortcomings of the existing method (Lotfi et al.: Applied Mathematical Modelling 33, 3151–3156, 2009) are pointed out and to overcome the limitations as well as to resolve the shortcomings, a new method proposed by Kumar et al. (Applied Mathematical Modelling 35, 817–823, 2011) is presented to find the nonnegative fuzzy optimal solution of fully fuzzy linear programming problems with equality constraints.

Chapter 3 presents the limitations of the method, presented in Chap. 2 and to overcome these limitations a method proposed by Kaur and Kumar (Applied Intelligence 37, 145–154, 2012) is presented for solving fully fuzzy linear programming problems with equality constraints. To show the application of the method presented in this chapter, a real-life problem, which cannot be solved using the method presented in Chap. 2 is solved by using the method presented in this chapter.

Chapter 4 describes that to the best of our knowledge, till now none has defined the product of such *LR* fuzzy numbers or *LR* flat fuzzy numbers that are neither nonnegative nor non-positive. Due to the nonexistence of such product, there was no method in the literature for solving such fully fuzzy linear programming problems in which some or all the parameters are represented by such *LR* fuzzy numbers or *LR* flat fuzzy numbers that are neither nonnegative nor non-positive. In this chapter, the product of such fuzzy numbers, proposed by Kaur and Kumar (Applied Mathematical Modelling 37, 7142–7153, 2013) is presented and also the limitations of the method, presented in Chap. 3, are pointed out. To overcome the limitations of the method, presented in Chap. 3, a method proposed by Kaur and Kumar (Applied Mathematical Modelling 37, 7142–7153, 2013) is presented to find the fuzzy optimal solution of fully fuzzy linear programming problems with equality constraints.

In Chap. 5, limitations of the existing methods for solving fuzzy linear programming problems and fully fuzzy linear programming problems with inequality constraints are pointed out. To overcome the limitations of the existing methods, two methods, proposed by Kumar and Kaur (Journal of Intelligent and Fuzzy Systems 26, 337–344, 2014) are presented, for solving fully fuzzy linear programming problems with inequality constraints. The advantages of the methods, presented in this chapter, over the existing methods are also discussed.

In Chap. 6, it is shown that the fuzzy optimal value, obtained by using the method presented in Chap. 4 is not necessarily a unique fuzzy number. So, it does not conform to the uniqueness property of fuzzy optimal value. To overcome this limitation of the method, presented in Chap. 4, a method proposed by Kaur and Kumar (Control and Cybernetics 41, 171–182, 2012; Journal of Optimization Theory and Applications 156, 529–534, 2013) is presented for solving fully fuzzy linear programming problems with equality constraints.

Finally, in Chap. 7, based on the present study, the future work has been suggested.

# Chapter 1
# State of the Art

Linear programming is a quantitative tool for an optimal allocation of limited resources amongst competing activities. It is perhaps the most popular operations research technique with applications in several functional areas of business such as production, finance, marketing, distribution, advertising and so forth [1–4].

The conventional linear programming deals with crisp parameters. However, in most of the real world applications, the nature of the parameters of the decision making problems is generally imprecise. For example, waste generation rate within a city is related to many socio-economic and environmental factors, and exhibits uncertain and dynamic features; the efficiency of a municipal waste water treatment plant is affected by waste water flow rate, and is uncertain in nature; regional air quality is mainly influenced by air pollutant emissions within this area, which also present uncertain characteristics [5]. Such uncertainties can lead to increased complexities in the related optimization efforts. Simply ignoring these uncertainties is considered undesired as it may result in inferior or wrong decisions [6–8]. Therefore, inexact linear programming methods are desired under uncertainty.

In the past decades, a large number of inexact linear programming methods have been developed to deal with various uncertainties. These methods were mainly classified into three categories, namely stochastic, interval and fuzzy linear programming [5, 7, 9–36].

Stochastic linear programming, derived from probability theory, could deal with various probabilistic uncertainties; however, the increased data requirements for specifying the parameters' probability distributions might affect their practical applicability [37].

Interval linear programming, based on interval analysis, was proved to be an effective approach in dealing with uncertainties. Interval linear programming did not require distributional information and would not lead to complicated intermediate models. However, it was to be noted that the outputs of Interval linear programming were with lower and upper bounds, and thus could not reflect the distribution of uncertainty within the lower and upper bounds [38].

© Springer International Publishing Switzerland 2016
J. Kaur and A. Kumar, *An Introduction to Fuzzy Linear
Programming Problems*, Studies in Fuzziness and Soft Computing 340,
DOI 10.1007/978-3-319-31274-3_1

Fuzzy sets theory and the related possibility theory, which were designed to supplement the interpretation of linguistic or measured uncertainties for real-world random phenomena, provided powerful tools for dealing with imprecise information [39].

Although fuzziness offered a weaker (less informative) indication of uncertainty than randomness, fuzzy sets theory and possibility theory did not require sufficient precise data to quantify imprecise information. Furthermore, fuzzy sets were more accurate than interval numbers, and reflected more detailed information for system uncertainties.

Fuzzy set theory has been extensively employed in linear programming. The main objective in fuzzy linear programming is to find the best solution possible with imprecise, vague, uncertain or incomplete information. There are many sources of imprecision in fuzzy linear programming. The sources of imprecision in fuzzy linear programming vary. For example, sometimes constraint satisfaction limits are vague and other times coefficient variables are not known precisely.

The research on fuzzy linear programming has risen highly since Bellman and Zadeh [40] proposed the concept of decision making in fuzzy environment. Zimmermann [41] introduced the first formulation of fuzzy linear programming to address the impreciseness and vagueness of the parameters in linear programming problems with fuzzy constraints and objective functions.

There are generally four fuzzy linear programming classifications in the literature:

1. Zimmerman [42] has classified fuzzy linear programming problems into two categories: symmetrical and non symmetrical models. In a symmetrical fuzzy decision there is no difference between the weight of the objectives and constraints while in the asymmetrical fuzzy decision, the objectives and constraints are not equally important and have different weights [43].
2. Leung [44] has classified fuzzy linear programming problems into four categories: (1) a precise objective and fuzzy constraints; (2) a fuzzy objective and precise constraints; (3) a fuzzy objective and fuzzy constraints; and (4) robust programming.
3. Luhandjula [45] has classified fuzzy linear programming problems into three categories: (1) flexible programming; (2) mathematical programming with fuzzy parameters; and (3) fuzzy stochastic programming.
4. Inuiguchi et al. [46] have classified fuzzy linear programming problems into six categories: (1) flexible programming; (2) possibilistic programming; (3) possibilistic linear programming using fuzzy max; (4) robust programming; (5) possibilistic programming with fuzzy preference relations; and (6) possibilistic linear programming with fuzzy goals.

A detailed review and classification of fuzzy linear programs and their solution approaches can be found in Baykasoglu and Gocken [47].

In the literature, most of the fuzzy linear programming models deal with the fuzziness related to the objective function(s)' coefficients, technological coefficients and right hand side values of the constraints. Because, available information on these parameters may not be precise and/or the decision maker(s) generally may not precisely know the values of these parameters. For example, Zhang et al. [48] proposed a fuzzy linear programming with fuzzy numbers for the coefficients of objective functions. They introduced a number of optimal solutions for the fuzzy linear programming problems and developed a number of theorems for converting the fuzzy linear programming problems into multi-objective optimization problems.

Stanciulescu et al. [49] proposed a fuzzy linear programming model with fuzzy coefficients for the objectives and the constraints. He used fuzzy decision variables with a joint membership function instead of crisp decision variables and linked the decision variables together to sum them up to a constant. He considered lower-bounded fuzzy decision variables that set up the lower bounds of the decision variables. He then generalized the method to lower-upper-bounded fuzzy decision variables that set up also the upper bounds of the decision variables.

Delgado et al. [12] studied a general model for fuzzy linear programming problems which simultaneously involved in the constraints set both fuzzy numbers and fuzzy constraints. Rommelfanger [50, 51] presented an interactive method for solving a multi-criteria linear program, where coefficients of the objective functions and/or of the constraints were expressed as $LR$ fuzzy numbers. Jimenez et al. [26] proposed an interactive resolution method for the fuzzy linear programming model with all coefficients were fuzzy numbers.

Mahdavi-Amiri and Nasseri [52] proposed a fuzzy linear programming model where a linear ranking function was used to rank order trapezoidal fuzzy numbers. They established the dual problem of the linear programming problem with trapezoidal fuzzy variables and deduced some duality results to solve the fuzzy linear programming problem directly with the primal simplex tableau.

Lodwick and Jamison [53] developed the theory underlying fuzzy, possibilistic, and mixed fuzzy/possibilistic optimization and demonstrated the appropriate use of distinct solution methods associated with each type of optimization dependent on the semantics of the problem. Lodwick and Untiedt [54] also discussed the concepts of fuzzy and possibilistic optimizations problems. Mahdavi-Amiri and Nasseri [35] developed some methods for solving fuzzy linear programming problems by introducing and solving certain auxiliary problems. They apply a linear ranking function to order trapezoidal fuzzy numbers and deduce some duality results by establishing the dual problem of the linear programming problem with trapezoidal fuzzy variables.

Rommelfanger [55] showed that both the probability distributions and fuzzy sets should be used in parallel or in combination, to model imprecise data dependent on the real situation. Hop [20] presented a model to measure attainment values of fuzzy numbers/fuzzy stochastic variables and used these new measures to convert the fuzzy linear programming problem or the fuzzy stochastic linear programming problem into the corresponding deterministic linear programming problem.

Ghodousian and Khorram [56] studied the new linear objective function optimization with respect to the fuzzy relational inequalities defined by max-min composition in which fuzzy inequality replaces ordinary inequality in the constraints. They showed that their method attains the optimal points that are better solutions than those resulting from the resolution of the similar problems with ordinary inequality constraints.

Tan et al. [57] developed a fuzzy linear programming extension of the general life cycle model using a concise and consistent linear model that makes identification of the optimal solution straightforward. Wu [58] derived the optimality conditions for fuzzy linear programming problems by proposing two solution concepts based on similar solution concept, called the non dominated solution, in the multi objective programming problem.

Gupta and Mehlawat [59] studied a pair of fuzzy primal-dual linear programming problems and calculated duality results using an aspiration level approach. Their approach is particularly important for fuzzy linear programming where the primal and dual objective values may not be bounded. Chen and Ko [1], Inuiguchi and Ramik [60] and Peidro et al. citepeidro have developed a number of fuzzy linear programming models to solve problems ranging from supply chain management to product development.

Ebrahimnejad [61] introduced a new primal-dual algorithm for solving fuzzy linear programming problems by using the duality results proposed by Mahdavi-Amiri and Nasseri [35]. Ebrahimnejad [62] has also generalized the concept of sensitivity analysis in fuzzy linear programming problems by applying fuzzy simplex algorithms and using the general linear ranking functions on fuzzy numbers.

Ebrahimnejad [63] showed that the method proposed by Ganesan and Veermani [64] stops in a finite number of iterations and proposed a revised version of their method that was more efficient and robust in practice. He also proved the absence of degeneracy and showed that if an fuzzy linear programming problem has a fuzzy feasible solution, it also has a fuzzy basic feasible solution and if an fuzzy linear programming problem has an optimal fuzzy solution, it also has an optimal fuzzy basic solution.

However, most of these methods were just applied to deal with fuzzy parameters or relationships. Few of them focused on solutions expressed as fuzzy sets in fuzzy linear programming problems. In real-world management problems, fuzzy solutions can provide some distributional information, and thus would be more attractive to decision makers.

In fact, decision variables of a fuzzy linear programming problem where all of the parameters are stated as fuzzy should also be considered as fuzzy numbers. Because, the fuzzy characteristic of the decision may be partially lost and the decision making process is limited with the crisp solutions when the decision variables oh is problem are crisp [11]. For this reason, instead of crisp solutions, obtaining fuzzy efficient solutions which provide ranges of flexibility to decision maker(s) seems more impressive in fully uncertain environments [17]. In other words, presenting the reasonable range of values for the decision variables can be comparatively better than the currently available crisp solutions [34].

In addition, fuzzy solutions may serve the regions containing potential satisfactory solutions around the optimal solutions to the decision maker(s). Thus, the final decisions can be made by the decision maker(s) as crisp ones. By taking the decision variables as fuzzy, the choice of the crisp decisions among the fuzzy solutions will also be supported [49]. Furthermore, it was emphasized by Xiaozhong et al. [65] that it is possible to encounter some optimization problems such as the cardinality of optimal solutions and the number of fairly superior solutions for a fuzzy linear program.

The concept of fuzzy linear programming with fuzzy decision variables was first introduced by Tanaka and Asai [66]. The fuzzy linear programming problems with fuzzy parameters can further be broadly classified into seven groups depending on the fuzziness of decision variables:

1. The fuzzy linear programming problems in this group involve fuzzy numbers for the decision variables and the right hand side of the constraints (e.g., Mahdavi-Amiri and Nasseri [35]).
2. The fuzzy linear programming problems in this group involve fuzzy numbers for the coefficients of the decision variables in the objective function (e.g., Wu [58]).
3. The fuzzy linear programming problems in this group involve fuzzy numbers for the coefficients of the decision variables in the constraints and the right hand side of the constraints (e.g., Xinwang [67]).
4. The fuzzy linear programming problems in this group involve fuzzy numbers for the decision variables, the coefficients of the decision variables in the objective function and the right hand side of the constraints (e.g., Ganesan and Veeramani [64]).
5. The fuzzy linear programming problems in this group involve fuzzy numbers for the coefficients of the decision variables in the objective function, the coefficients of the decision variables in the constraints and the right hand side of the constraints (e.g., Mahdavi-Amiri and Nasseri [52], Hatami-Marbini and Tavana [68], Hatami-Marbini et al. [69]).
6. The fuzzy linear programming problems in this group involve fuzzy numbers for the coefficients of the decision variables in the decision variables, the coefficients of the decision variables in the constraints and the right hand side of the constraints (e.g., Saati et al. [70]).
7. The fuzzy linear programming problems in this group, so called fully fuzzy linear programming problems, involve fuzzy numbers in the decision variables, the coefficients of the decision variables in the objective function, the coefficients of the decision variables in the constraints and the right hand side of the constraints (e.g., Lotfi et al. [33]).

Fully fuzzy linear programming problems have been an attractive topic for the researchers in recent years. However, the literature on fully fuzzy linear programming is not rich as there are a few studies available on this research topic.

Tanaka and Asai [71] solved fully fuzzy linear programming problems in their paper. They handled fuzzy linear programming problems with fuzzy satisfaction criteria and fuzzy parameters. The goals and constraints are accepted as identical concepts. The fuzzy linear programming problem is converted into crisp problem as in Tanaka and Asai [66].

Tanaka et al. [72] studied on obtaining fuzzy decision to fuzzy decision making problems using possibility distributions of fuzzy decision variables. In their study, Tanaka et al. solved fuzzy linear programming problems with fuzzy right hand values and fuzzy decision variables. For solving the fuzzy problem, they used possibility distributions of fuzzy decision variables and transformed fuzzy problems into crisp problem.

Buckley and Feuring [9] proposed a solution method for fully fuzzy linear programming problems. All of the parameters and decision variables are defined as triangular fuzzy numbers. The fuzzy linear programming problem is transformed into a multi-objective fuzzy linear programming problem. For example, for a maximizing problem it is tried to maximize the midpoint of the fuzzy objective, minimize the area between the midpoint and the minimum point of the membership function of the fuzzy objective and maximize the area between the midpoint and the maximum point of the membership function of fuzzy objective. The fuzzy constraints are handled using fuzzy ranking methods. Fuzzy flexible programming problem is used to find the whole un-dominated set of the multi objective fuzzy linear programming problem.

Buckley et al. [73] solved multi-objective fully fuzzified linear programming problems in their study. All the parameters and variables are defined as triangular fuzzy numbers. The same solution procedure proposed in Buckley and Feuring [9] is used for the multi-objective fully fuzzified linear programming problems. The multi-objective fully fuzzified linear programming problem is changed into a single objective fuzzy linear programming problem and then solved by using the proposed solution procedure. An evolutionary algorithm is used to generate un-dominated solutions.

Tsakiris and Spiliotis [74] presented a methodology for solving the problem of water allocation to various users under uncertainty. In one instance of their study, they defined decision variables as fuzzy. In the study of Tsakiris and Spiliotis, the water allocation problem with fuzzy decision variables is solved using the method which was proposed by Tanaka et al. [72].

Hashemi et al. [19] proposed a solution method for fully fuzzified linear programming problems in which all parameters and decision variables are defined as symmetric fuzzy numbers. The solution procedure is constructed on a ranking method which is based on the comparison of mean and standard deviation of fuzzy numbers. In fuzzy arithmetic based on Zadeh's extension principle, the shape of $LR$ fuzzy number is not preserved. So, Hashemi et al. used the new fuzzy arithmetic operations on symmetric fuzzy numbers introduced by Nasrabadi and Nasrabadi [75]. Hashemi et al. [19] proposed a two phase approach for the solution of the fully fuzzified linear programming problems. In the first phase, the possibilistic mean value of fuzzy objective function is tried to be maximized and a set of feasible solutions are obtained. In the second phase, the standard deviation of the original fuzzy objective function is tried to be minimized by considering all basic feasible solutions obtained at the end of the first phase.

Allahviranloo et al. [76] proposed a method to solve fully fuzzy linear programming problems. They defined all of the parameters of the problem and the decision variables as triangular fuzzy numbers. The fully fuzzy linear programming problem is defuzzified using a linear ranking function and the crisp equivalent of the fully fuzzy linear programming problem is obtained. Ebrahimnejad and Nasseri [77] developed the complementary slackness theorem for solving fuzzy linear programming problems with fuzzy parameters.

A fully fuzzy linear programming problem was transformed into the two crisp linear problems by Lotfi et al. [33] that are maximization problem for the core and minimization problem for the margin. The proposed method is not an exact one, just an approximation technique since it relies on the approximation of the fuzzy numbers to the nearest symmetric triangular fuzzy numbers. Kumar et al. [78] considered this problem and presented a new method for solving FFLP problems. However this method is limited to a certain class of fuzzy numbers, i.e. triangular fuzzy numbers and needed some corrections made by Najafi and Edalatpanah [79].

Baykasoglu and Gocken [80] developed a direct solution approach for solving linear programming problems with fuzzy decision variables based on integrating a fuzzy ranking method in to a meta-heuristic algorithm which aims to rank the objective function values and determine the feasibility degrees of the constraints.

Ezzati et al. [81] pointed out that recently, two methods have been introduced to solve the FFLP problems by Lotfi et al. [33] and Kumar et al. [78]. In the first method [33], the parameters of fully fuzzy linear programming problem have been approximated to the nearest symmetric triangular fuzzy numbers. After that, a fuzzy optimal approximation solution has been achieved by solving a multi objective linear programming problem. The shortcoming exists of it is that the optimal solution of fully fuzzy linear programming problem is not exact. So, it is not reliable solution for decision maker. In the second method [78], an exact optimal solution is achieved using a linear ranking function. In this method, the linear ranking function has been used to convert the fuzzy objective function to the crisp objective function. The shortcoming exists of it is that the fuzziness of objective function has been neglected by the linear ranking function. Ezzati et al. [81] proposed a new lexicographic ordering on triangular fuzzy numbers and using it proposed a method to solve fully fuzzy linear programming problem. To find the fuzzy optimal solution of a fully fuzzy linear programming problem, Ezzati et al. [81] firstly transformed the fully fuzzy linear programming problem into a multi objective linear programming problem with three objective functions. Then, used lexicographic ordering on triangular fuzzy numbers, proposed by themselves, to find a lexicographic optimal solution of multi objective linear programming problem.

Ezzati et al. [81] also claimed that the fuzzy optimal solution of fully fuzzy linear programming problems with inequality constraints can also be obtained by the same algorithm by transforming it into fully fuzzy linear programming problems with equality constraints. Bhardwaj and Kumar [82] pointed out that proved that the fully fuzzy linear programming problems with inequality constraints cannot be transformed into fully fuzzy linear programming problems with equality constraints and hence, the algorithm, proposed by Ezzati et al. [81] to find the fuzzy optimal solution of fully fuzzy linear programming problems with equality constraints, cannot be used for finding the fuzzy optimal solution of fully fuzzy linear programming problems with inequality constraints.

Khan et al. [83] claimed that there is no method in the literature to find the fuzzy optimal solution of a fully fuzzy linear programming problem without converting it into crisp linear programming problem, and proposed a technique for the same. Khan et al. [83] also introduced the dual of fully fuzzy linear programming problem. Bhardwaj and Kumar [82] shown that Khan et al. [83] have used some such assumptions, in their proposed technique as well as in writing the dual of fully fuzzy linear programming problem, which are mathematically incorrect.

Fan et al. [84] proposed generalized fuzzy linear programming method based on the idea of design of experiment and the interactive algorithm proposed by Huang et al. [21–25] for interval parameter linear programming problems. This method is implemented through (i) discretizing membership grade of fuzzy parameters in fully fuzzy linear programming into a finite number of alpha-cut levels, (ii) converting the fully fuzzy linear programming model into an interval linear programming submodel under every alpha-cut level; (iii) solving the interval linear programming submodels through an interactive algorithm and obtaining the associated interval solutions, (iv) acquiring the membership functions of fuzzy solutions through statistical regression methods. Kumar et al. [85] pointed out that the solution obtained by the algorithm proposed by Fan et al. [84] is not necessarily a fuzzy number.

All the above mentioned existing methods for solving fully fuzzy linear programming problems are proposed by considering the assumption that all the parameters and decision variables are non-negative triangular/trapezoidal/$LR$ flat fuzzy numbers. Kumar et al. [78] was the first one who tried to overcome this limitation of the existing methods by proposing a method to find the fuzzy optimal solution of such fully fuzzy linear programming problems with equality constraints in which the decision variables are represented by non-negative triangular fuzzy numbers whereas all the other parameters are represented by unrestricted triangular fuzzy numbers. Kumar and Kaur [86] proposed a method to find the fuzzy optimal solution of same type of fully fuzzy linear programming problems with inequality constraints. Najafi and Edalatpanah [79] suggested some corrections in the method proposed by Kumar et al. [78].

Kaur and Kumar [87] pointed out the method, proposed by Kumar et al. [78] cannot be used to find the fuzzy optimal solution of such fully fuzzy linear programming problems in which decision variables are represented by unrestricted fuzzy numbers. To overcome this limitation of the existing method [78], Kaur and Kumar [87] proposed a method to find the fuzzy optimal solution of such fully fuzzy linear programming problems with equality constraints in which the decision variables are represented by unrestricted trapezoidal fuzzy numbers whereas all the other parameters are represented by non-negative trapezoidal fuzzy numbers. Kaur and Kumar [87] also shown that their method can also be used to find the fuzzy optimal solution of fully fuzzy linear programming problems considered by Kumar et al. [78].

Kaur and Kumar [88] pointed out that the methods, proposed by Kaur and Kumar [87], cannot be used to find the fuzzy optimal solution of such fully fuzzy linear programming problems in which all the parameters and decision variables are represented by unrestricted fuzzy numbers. To overcome this limitation of the method, proposed by Kaur and Kumar [87], Kaur and Kumar [88] proposed the product of unrestricted $LR$ flat fuzzy numbers and using it, Kaur and Kumar [88] proposed a method, named as Mehar's method, to find the fuzzy optimal solution of such fully fuzzy linear programming problems in which all the parameters and decision variables are represented by unrestricted $LR$ flat fuzzy numbers.

Kaur and Kumar [89, 90] pointed out that the fuzzy optimal value of a fuzzy linear programming problem should be a unique fuzzy numbers. However, on solving a fully fuzzy linear programming problem by using the existing methods, the obtained fuzzy optimal value is not necessarily a unique fuzzy number. To resolve this drawback of the existing methods, Kaur and Kumar [89, 90] proposed a method to find the unique optimal value of such linear programming problems with equality constraints in which decision variables are represented by real numbers whereas all other parameters are represented by unrestricted $LR$ flat fuzzy numbers. The same method was extended by Kaur [91] for finding the fuzzy optimal value of such fully fuzzy linear programming problems with equality constraints in which all the parameters and decision variables are represented by unrestricted $LR$ flat fuzzy numbers.

As the literature in this field is voluminous and cannot be cited in full but luckily enough there exist in the literature two edited volumes devoted to fuzzy optimization [92, 93] which are sources of information on virtually all newer and more relevant developments in the respective field. Other more extensive surveys on fuzzy optimization and fuzzy mathematical programming, the interested reader is referred to, e.g., [94–105].

The aim of this monograph is to present the methods for solving fully fuzzy linear programming problems, proposed by the authors of this monograph, at one place. This book presents the preliminaries, fundamentals, methods to find the optimal solution, optimality of fully fuzzy linear programming problems and the comparative study as well.

# References

1. Chen, L.H., Ko, W.C.: Fuzzy linear programming models for new product design using QFD with FMEA. Appl. Math. Model. **33**, 633–647 (2009)
2. Amin, S.H., Razmi, J., Zhang, G.: Supplier selection and order allocation based on fuzzy SWOT analysis and fuzzy linear programming. Expert Syst. Appl. **38**, 334–342 (2011)
3. Peidro, D., Mula, J., Jimenez, M., Botella, M.: A fuzzy linear programming based approach for tactical supply chain planning in an uncertainty environment. Eur. J. Oper. Res. **205**, 65–80 (2010)
4. Rong, A., Lahdelma, R.: Fuzzy chance constrained linear programming model for optimizing the scrap charge in steel production. Eur. J. Oper. Res. **186**, 953–964 (2008)
5. Fan, Y.R., Huang, G.H.: A robust two-step method for solving interval linear programming problems within an environmental management context. J. Environ. Inf. **19**, 1–9 (2012)
6. Ozdemir, M.S., Saaty, T.L.: The unknown in decision making what to do about it?. Eur. J. Oper. Res. **181**, 1434–1463 (2006)

7. Ruszcynski, A.: Decomposition methods in stochastic programming. Math. Prog. **79**, 333–353 (1997)
8. Young, R.A.: Uncertainty and the Environment: Implications for Decision Making and Environmental Policy. Edward Elgar, Cheltenham (2001)
9. Buckley, J., Feuring, T.: Evolutionary algorithm solution to fuzzy problems: fuzzy linear programming. Fuzzy Sets Syst. **109**, 35–53 (2000)
10. Chanas, S.: The use of parametric programming in fuzzy linear programming. Fuzzy Sets Syst. **11**, 229–241 (1983)
11. Dehghan, M., Hashemi, B., Ghatee, M.: Computational methods for solving fully fuzzy linear systems. Appl. Math. Comput. **179**, 328–343 (2006)
12. Delgado, M., Verdegay, J.L., Vila, M.A.: A general model for fuzzy linear programming. Fuzzy Sets Syst. **29**, 21–29 (1989)
13. Delgado, M., Herrera, F., Verdegay, J.L., Vila, M.A.: Post-optimality analysis on the membership functions of a fuzzy linear programming problems. Fuzzy Sets Syst. **53**, 289–297 (1993)
14. Fan, Y.R., Huang, G.H., Li, Y.P., Cao, M.F., Cheng, G.H.: A fuzzy linear programming approach for municipal solid-waste management under uncertainty. Eng. Optim. **41**, 1081–1101 (2009)
15. Fan, Y.R., Huang, G.H., Veawab, A.: A generalized fuzzy linear programming approach for environmental management problem under uncertainty. J. Air Waste Manage. Assoc. **62**, 72–86 (2012)
16. Fan, Y.R., Huang, G.H., Guo, P., Yang, A.L.: Inexact two-stage stochastic partial programming: application to water resources management under uncertainty. Stoch. Env. Res. Risk Assess. **26**, 281–293 (2012)
17. Fan, Y.R., Huang, G.H., Li, Y.P.: Robust interval linear programming for environmental decision making under uncertainty. Eng. Optim. **44**, 1321–1336 (2012)
18. Guo, L., Li, Y.P., Huang, G.H., Wang, X.W., Dai, C.: Development of an interval-based evacuation management model in response to nuclear-power plant accident. J. Environ. Inf. **20**, 58–66 (2012)
19. Hashemi, S.M., Modarres, M., Nasrabadi, E., Nasrabadi, M.M.: Fully fuzzified linear programming, solution and duality. J. Intell. Fuzzy Syst. **17**, 253–261 (2006)
20. Hop, N.V.: Solving fuzzy (stochastic) linear programming problems using superiority and inferiority measures. Inf. Sci. **177**, 1977–1991 (2007)
21. Huang, G.H.: IPWM: an interval parameter water quality model. Eng. Optim. **26**, 79–103 (1996)
22. Huang, G.H.: A hybrid inexact-stochastic water management model. Eur. J. Oper. Res. **107**, 137–158 (1998)
23. Huang, G.H., Baetz, B.W., Patry, G.G.: A grey linear programming approach for municipal solid waste management planning under uncertainty. Civ. Eng. Syst. **9**, 319–335 (1992)
24. Huang, G.H., Baetz, B.W., Patry, G.G.: A grey fuzzy linear programming approach for municipal solid waste management planning under uncertainty. Civ. Eng. Syst. **10**, 123–146 (1993)
25. Huang, G.H., Baetz, B.W., Patry, G.G.: Grey fuzzy integer programming: An application to waste management planning under uncertainty. Eur. J. Oper. Res. **83**, 594–620 (1995)
26. Jimenez, M., Arenas, M., Bilbao, A., Rodrguez, M.V.: Linear programming with fuzzy parameters: an interactive method resolution. Eur. J. Oper. Res. **177**, 1599–1609 (2007)
27. Katagiri, H., Sakawa, M.: Interactive multi objective fuzzy random programming through the level set-based probability model. Inf. Sci. **181**, 1641–1650 (2011)
28. Lee, Y.W., Bogardi, I., Stansbury, J.: Fuzzy decision making in dredged-material management. ASCE J. Environ. Eng. **117**, 614–628 (1991)
29. Leon, T., Vercher, E.: Solving a class of fuzzy linear programs by using semi-infinite programming techniques. Fuzzy Sets Syst. **146**, 235–252 (2004)
30. Li, G.: Fuzzy goal programming - A parametric approach. Inf. Sci. **195**, 287–295 (2012)
31. Li, Y.P., Huang, G.H.: Fuzzy stochastic based violation analysis method for planning water resources management systems with uncertain information. Inf. Sci. **179**, 4261–4276 (2009)

32. Li, Y.P., Huang, G.H., Guo, P., Yang, Z.F., Nie, S.L.: A dual-interval vertex analysis method and its application to environmental decision making under uncertainty. Eur. J. Oper. Res. **200**, 536–550 (2010)
33. Lotfi, F.H., Allahviranloo, T., Jondabeha, M.A., Alizadeh, L.: Solving a fully fuzzy linear programming using lexicography method and fuzzy approximate solution. Appl. Math. Model. **33**, 3151–3156 (2009)
34. Maleki, H.R., Tata, M., Mashinchi, M.: Linear programming with fuzzy variables. Fuzzy Sets Syst. **109**, 21–33 (2000)
35. Mahdavi-Amiri, N., Nasseri, S.H.: Duality results and a dual simplex method for linear programming problems with trapezoidal fuzzy variables. Fuzzy Sets Syst. **158**, 1961–1978 (2007)
36. Nguyen, V.H.: Solving linear programming problems under fuzziness and randomness environment using attainment values. Inf. Sci. **177**, 2971–2984 (2007)
37. Nie, X.H., Huang, G.H., Li, Y.P., Liu, L.: IFRP: a hybrid interval-parameter fuzzy robust programming approach for waste management planning under uncertainty. J. Environ. Manage. **84**, 1–11 (2007)
38. Srivastava, A.K., Nema, A.K.: Fuzzy parametric programming model for integrated solid waste management under uncertainty. J. Environ. Eng. **137**, 69–83 (2011)
39. Zadeh, L.A.: Fuzzy sets. Inf. Control **8**, 338–353 (1965)
40. Bellman, R.E., Zadeh, L.A.: Decision making in a fuzzy environment. Manage. Sci. **17**, 141–164 (1970)
41. Zimmermann, H.J.: Fuzzy programming and linear programming with several objective functions. Fuzzy Sets Syst. **1**, 45–55 (1978)
42. Zimmermann, H.J.: Fuzzy Sets. Decision Making and Expert Systems. Kluwer Academic Publishers, MA (1987)
43. Amid, A., Ghodsypour, S.H., O'Brien, C.: Fuzzy multi objective linear model for supplier selection in a supply chain. Int. J. Prod. Econ. **104**, 394–407 (2006)
44. Leung, Y.: Spatial Analysis and Planning Under Imprecision. North-Holland, Amsterdam (1988)
45. Luhandjula, M.K.: Fuzzy optimization: an appraisal. Fuzzy Sets Syst. **30**, 257–282 (1989)
46. Inuiguchi, M., Ichihashi, H., Tanaka, H.: Fuzzy programming: a survey of recent developments. In: Slowinski, R., Teghem, J. (eds.) Stochastic Versus Fuzzy Approaches to Multi Objective Mathematical Programming Under Uncertainty, pp. 45–68. Kluwer Academic Publishers, Dordrecht (1990)
47. Baykasoglu, A., Gocken, T.: A review and classification of fuzzy mathematical programs. J. Intell. Fuzzy Syst.: Appl. Eng. Technol. **19**, 205–229 (2008)
48. Zhang, G., Wu, Y.H., Remias, M., Lu, J.: Formulation of fuzzy linear programming problems as four-objective constrained optimization problems. Appl. Math. Comput. **139**, 383–399 (2003)
49. Stanciulescu, C., Fortemps, Ph., Installe, M., Wertz, V.: Multiobjective fuzzy linear programming problems with fuzzy decision variables. Eur. J. Oper. Res. **149**, 654–675 (2003)
50. Rommelfanger, H.: Interactive decision making in fuzzy linear optimization problems. Eur. J. Oper. Res. **41**, 210–217 (1989)
51. Rommelfanger, H.: Fuzzy linear programming and applications. Eur. J. Oper. Res. **92**, 512–527 (1996)
52. Mahdavi-Amiri, N., Nasseri, S.H.: Duality in fuzzy number linear programming by the use of a certain linear ranking function. Appl. Math. Comput. **180**, 206–216 (2006)
53. Lodwick, W.A., Jamison, K.D.: Theoretical and semantic distinctions of fuzzy, possibilistic, and mixed fuzzy/possibilistic optimization. Fuzzy Sets Syst. **158**, 1861–1872 (2007)
54. Lodwick, W.A., Untiedt, E.: Introduction to Fuzzy and Possibilistic Optimization in Chapter 1: Fuzzy Optimization: Recent Developments and Applications. In: W.A. Lodwick and J. Kacprzyk (eds.). Springer, New York (2010)
55. Rommelfanger, H.: A general concept for solving linear multicriteria programming problems with crisp, fuzzy or stochastic values. Fuzzy Sets Syst. **158**, 1892–1904 (2007)

56. Ghodousian, A., Khorram, E.: Fuzzy linear optimization in the presence of the fuzzy relation inequality constraints with max-min composition. Inf. Sci. **178**, 501–519 (2008)
57. Tan, R.R., Culaba, A.B., Aviso, K.B.: A fuzzy linear programming extension of the general matrix-based life cycle model. J. Cleaner Prod. **16**, 1358–1367 (2008)
58. Wu, H.C.: Optimality conditions for linear programming problems with fuzzy coefficients. Comput. Math. Appl. **55**, 2807–2822 (2008)
59. Gupta, P., Mehlawat, M.K.: Bector-Chandra type duality in fuzzy linear programming with exponential membership functions. Fuzzy Sets Syst. **160**, 3290–3308 (2009)
60. Inuiguchi, M., Ramik, J.: Possibilistic linear programming: A brief review of fuzzy mathematical programming and a comparison with stochastic programming in portfolio selection problem. Fuzzy Sets Syst. **111**, 3–28 (2000)
61. Ebrahimnejad, A., Nasseri, S.H., Lotfi, F.H., Soltanifar, M.: A primal-dual method for linear programming problems with fuzzy variables. Eur. J. Oper. Res. **4**, 189–209 (2010)
62. Ebrahimnejad, A.: Sensitivity analysis in fuzzy number linear programming problems. Math. Comput. Model. **53**, 1878–1888 (2011)
63. Ebrahimnejad, A.: Some new results in linear programs with trapezoidal fuzzy numbers: finite convergence of the Ganesan and Veeramani's method and a fuzzy revised simplex method. Appl. Math. Model. **35**, 4526–4540 (2011)
64. Ganesan, K., Veeramani, P.: Fuzzy linear programs with trapezoidal fuzzy numbers. Ann. Oper. Res. **143**, 305–315 (2006)
65. Xiaozhong, L., Yuyue, D., Fengehao, Z.: Fuzzy linear programming problems with fuzzy variables and fuzzy coefficients. J. Liaocheng Teachers College 86–90 (1998)
66. Tanaka, H., Asai, K.: Fuzzy solution in fuzzy linear programming problems. IEEE Trans. Syst. Man Cybern. **SMC-14** 325–328 (1984)
67. Xinwang, L.: Measuring the satisfaction of constraints in fuzzy linear programming. Fuzzy Sets Syst. **122**, 263–275 (2001)
68. Hatami-Marbini, A., Tavana, M.: An extension of the linear programming method with fuzzy parameters. Int. J. Math. Oper. Res. **3**, 44–55 (2011)
69. Hatami-Marbini, A., Agrell, P., Tavana, M., Emrouznejad, A.: A stepwise fuzzy linear programming model with possibility and necessity relation. J. Intell. Fuzzy Syst. **25**, 81–93 (2013)
70. Saati, S., Hatami-Marbini, A., Tavana, M., Hajiahkondi, E.: A two-fold linear programming model with fuzzy data. Int. J. Fuzzy Syst. Appl. **2**, 1–12 (2012)
71. Tanaka, H., Asai, K.: Fuzzy linear programming problems with fuzzy numbers. Fuzzy Sets Syst. **13**, 1–10 (1984)
72. Tanaka, H., Guo, P., Zimmermann, H.J.: Possibility distributions of fuzzy decision variables obtained from possibilistic linear programming problems. Fuzzy Sets Syst. **113**, 323–332 (2000)
73. Buckley, J.J., Feuring, T., Hayashi, Y.: Multi-objective fully fuzzified linear programming. Int. J. Uncertainty Fuzziness Knowl. Based Syst. **9**, 605–621 (2001)
74. Tsakiris, G., Spiliotis, M.: Fuzzy linear programming for problems of water allocation under uncertainty. Eur. Water **7**(8), 25–37 (2004)
75. Nasrabadi, M.M., Nasrabadi, E.: A mathematical programming approach to fuzzy linear regression analysis. Appl. Math. Comput. **155**, 873–881 (2004)
76. Allahviranloo, T., Lotfi, F.H., Kiasary, M.K., Kiani, N.A., Alizadeh, L.: Solving fully fuzzy linear programming problem by the ranking functhion. Appl. Math. Sci. **2**, 19–32 (2008)
77. Ebrahimnejad, A., Nasseri, S.H.: Using complementary slackness property to solve linear programming with fuzzy parameters. Fuzzy Inf. Eng. **1**, 233–245 (2009)
78. Kumar, A., Kaur, J., Singh, P.: A new method for solving fully fuzzy linear programming problems. Appl. Math. Model. **35**, 817–823 (2011)
79. Najafi, H.S., Edalatpanah, S.A.: A note on "A new method for solving fully fuzzy linear programming problems". Appl. Math. Model. **37**, 7865–7867 (2013)
80. Baykasoglu, A., Gocken, T.: A direct solution approach to fuzzy mathematical programs with fuzzy decision variables. Expert Syst. Appl. **39**, 1972–1978 (2012)

81. Ezzati, R., Khorram, E., Enayati, R.: A new algorithm to solve fully fuzzy linear programming problems using the MOLP problem. Appl. Math. Model. **39**, 3183–3193 (2015)
82. Bhardwaj, B., Kumar, A.: A note on the paper "A simplified novel technique for solving fully fuzzy linear programming problems". J. Optim. Theory Appl. **163**, 685–696 (2014)
83. Khan, I.U., Ahmad, T., Maan, N.: A simplified novel technique for solving fully fuzzy linear programming problems. J. Optim. Theory Appl. **159**, 536–546 (2013)
84. Fan, Y.R., Huang, G.H., Yang, A.L.: Generalized fuzzy linear programming for decision making under uncertainty: feasibility of fuzzy solutions and solving approach. Inf. Sci. **241**, 12–27 (2013)
85. Kumar, A., Appadoo, S.S., Bector, C.R.: A note on "Generalized fuzzy linear programming for decision making under uncertainty: feasibility of fuzzy solutions and solving approach". Inf. Sci. **266**, 226–227 (2014)
86. Kumar, A., Kaur, J.: Fuzzy optimal solution of fully fuzzy linear programming problems using ranking function. J. Intell. Fuzzy Syst. **26**, 337–344 (2014)
87. Kaur, J., Kumar, A.: Exact fuzzy optimal solution of fully fuzzy linear programming problems with unrestricted fuzzy variables. Appl. Intell. **37**, 145–154 (2012)
88. Kaur, J., Kumar, A.: Mehar's method for solving fully fuzzy linear programming problems with $LR$ fuzzy parameters. Appl. Math. Model. **37**, 7142–7153 (2013)
89. Kaur, J., Kumar, A.: A new method to find the unique fuzzy optimal value of fuzzy linear programming problems. J. Optim. Theory Appl. **156**, 529–534 (2013)
90. Kaur, J., Kumar, A.: Unique fuzzy optimal value of fully fuzzy linear programming problems. Control Cybern. **41**, 171–182 (2012)
91. Kaur, J.: Methods for Solving Linear Programming Problems with Fuzzy Parameters. Thapar University, Punjab (2012)
92. Kacprzyk, J., Orlovski, S.A.: Optimization Models Using Fuzzy Sets and Possibility Theory. D. Reidel Publishing Company, Dordrecht (1987)
93. Delgado, D., Kacprzyk, J., Verdegy, J.L., Villa, M.A.: Fuzzy Optimization: Recent Advances. Physica Verlag, Germany (1994)
94. Kacprzyk, J.: F 5.2 Optimization. In: Ruspini, E.H., Bonissone, P.P., Pedrycz, W. (eds.) Handbook of Fuzzy Computation, pp. F5.2:1-20. Institute of Physics Publishing Ltd, Bristol and Philadelphia (1998)
95. Kacprzyk, J., Orlovski, S.A.: Fuzzy optimization and mathematical programming: a brief introduction and survey. In: Kacprzyk, J., Orlovski, S.A. (eds.) Optimization Models Using Fuzzy Sets and Possibility Theory, pp. 50–72. Lancaster, Dordrecht (1987)
96. Fedrizzi, M., Kacprzyk, J., Verdegay, J.L.: A survey of fuzzy optimization and mathematical programming. In: Fedrizzi, M., Kacprzyk, J., Roubens, M. (eds.) Interactive Fuzzy Optimization, pp. 15–28. Springer, Heidelberg (1991)
97. Lodwick, W.A., Kacprzyk, J. (eds.): Fuzzy Optimization: Recent Advances and Applications. Springer, Heidelberg (2010)
98. Verdegay, J.L.: Progress on fuzzy mathematical programming: a personal perspective. Fuzzy Sets Syst. **281**, 219–226 (2015)
99. Fedrizzi, M., Kacprzyk, J., Roubens, M. (eds.): Interactive Fuzzy Optimization. Springer, Berlin (1991)
100. Kacprzyk, J.: Multistage Decision Making under Fuzziness. Verlag TUV, Rheinland, Koln (1983)
101. Kacprzyk, J. (ed.): Special issue on fuzzy sets and possibility theory in optimization models. Control Cybern. **4** (1984)
102. Lodwick, W.A., Jamison, K.D.: Theory and semantics for fuzzy and possibilistic optimization. In: Fuzzy Logic, Soft Computing and Computational Intelligence (Eleventh International Fuzzy Systems Association World Congress), July 28–31, vol. 2005, pp. 1805–1810. Beijing, China (2005)
103. Lodwick, W.A., Jamison, K.D.: The use of interval-valued probability measure in optimization under uncertainty for problems containing a mixture of possibilistic, probabilistic and interval uncertainty. In: Fundamentals of Fuzzy Logic and Soft Computing, 12th International Fuzzy

Systems Association World Congress, IFSA: Cancun, Mexico, June 18–21. Proceedings, vol. 2007, pp. 361–370 (2007)

104. Lodwick, W.A., Jamison, K.D.: Interval-valued probability in the analysis of problems containing a mixture of possibilistic, probabilistic, and interval uncertainty. Fuzzy Sets Syst. **159**, 2845–2858 (2008)

105. Lodwick, W.A., Neumaier, A., Newman, F.D.: Optimization under uncertainty: methods and applications in radiation therapy. In: Proceedings 10th IEEE International Conference on Fuzzy Systems 2001, vol. 3, pp. 1219–1222 (2001)

# Chapter 2
# Non-negative Fuzzy Optimal Solution of Fully Fuzzy Linear Programming Problems with Equality Constraints

Lotfi et al. [1] pointed out that there is no method in the literature for solving fully fuzzy linear programming problems with equality constraints and proposed a method for the same. In this chapter, the limitations and shortcoming of the existing method [1] are pointed out and to overcome the limitations as well as to resolve the shortcoming, Kumar et al.'s method [2] is presented to find the non-negative fuzzy optimal solution of fully fuzzy linear programming problems with equality constraints.

## 2.1  Preliminaries

In this section, some basic definitions and arithmetic operations for trapezoidal fuzzy numbers are presented [3].

### 2.1.1  Basic Definitions

In this section, some basic definitions are presented.

**Definition 2.1** Let $X$ be a classical set of objects. Then, the set of ordered pairs $\tilde{A} = \{(x, \mu_{\tilde{A}}(x)) : x \in X\}$, where $\mu_{\tilde{A}} : X \rightarrow [0, 1]$, is called a fuzzy set in $X$. The evaluation function $\mu_{\tilde{A}}(x)$ is called the membership function.

**Definition 2.2** Let $\tilde{A}$ be a fuzzy set in $X$ and $\lambda \in [0, 1]$ be a real number. Then, a classical set $A^{\lambda} = \{x \in X : \mu_{\tilde{A}}(x) \geq \lambda\}$ is called an $\lambda$ level set or $\lambda$-cut or parametric form of $\tilde{A}$.

---

The contents of this chapter are published in *Applied Mathematical Modelling* 35 (2011) 817–823.

© Springer International Publishing Switzerland 2016
J. Kaur and A. Kumar, *An Introduction to Fuzzy Linear Programming Problems*, Studies in Fuzziness and Soft Computing 340,
DOI 10.1007/978-3-319-31274-3_2

**Definition 2.3** A fuzzy set $\tilde{A} = \{(x, \mu_{\tilde{A}}(x)) : x \in X\}$ is called a normalized fuzzy set if and only if $\text{Supremum}_{x \in X}\{\mu_{\tilde{A}}(x)\} = 1$.

**Definition 2.4** A fuzzy set $\tilde{A}$ is called a convex fuzzy set if and only if $\mu_{\tilde{A}}(\alpha x_1 + (1 - \alpha)x_2) \geq \min\{\mu_{\tilde{A}}(x_1), \mu_{\tilde{A}}(x_2)\}$, $\forall x_1, x_2 \in X$, $\alpha \in [0, 1]$.

**Definition 2.5** A convex normalized fuzzy set $\tilde{A} = \{(x, \mu_{\tilde{A}}(x)) : x \in X\}$ is called a fuzzy number if and only if $\mu_{\tilde{A}}(x)$ is piecewise continuous in $X$.

**Definition 2.6** A fuzzy number $\tilde{A}$ is said to be a non-negative fuzzy number if and only if $\mu_{\tilde{A}}(x) = 0$, $\forall x < 0$.

**Definition 2.7** A fuzzy number $\tilde{A}$ defined on the universal set of real numbers $\mathbb{R}$, denoted as $\tilde{A} = (a, b, c, d)$, is said to be a trapezoidal fuzzy number if its membership function, $\mu_{\tilde{A}}(x)$, is given by

$$\mu_{\tilde{A}}(x) = \begin{cases} \frac{(x-a)}{(b-a)} & a \leq x < b \\ 1 & b \leq x \leq c \\ \frac{(x-d)}{(c-d)} & c < x \leq d \\ 0 & \text{otherwise} \end{cases}$$

**Definition 2.8** Let $\tilde{A} = (a, b, c, d)$ be a trapezoidal fuzzy number. Then, its $\lambda$-cut $A^\lambda$ is defined as follows:

$$A^\lambda = [a + (b - a)\lambda, d - (d - c)\lambda], \ 0 \leq \lambda \leq 1$$

**Definition 2.9** A trapezoidal fuzzy number $\tilde{A} = (a, b, c, d)$ is said to be symmetric trapezoidal fuzzy number if and only if $b - a = d - c$, otherwise $\tilde{A}$ is said to be an asymmetric trapezoidal fuzzy number.

A symmetric trapezoidal fuzzy number $\tilde{A} = (a, b, c, d)$ can be denoted as $(b, c, \sigma)$, where $[b, c]$ is the core and $\sigma = b - a = d - c$ is the spread of the symmetric trapezoidal fuzzy number $\tilde{A}$.

**Definition 2.10** A trapezoidal fuzzy number $\tilde{A} = (a, b, c, d)$ is said to be non-negative trapezoidal fuzzy number if and only if $a \geq 0$ and is said to be non-positive trapezoidal fuzzy number if and only if $a \leq 0$.

**Definition 2.11** A trapezoidal fuzzy number $\tilde{A} = (a, b, c, d)$ is said to be unrestricted trapezoidal fuzzy number if and only if $a$ is a real number.

**Definition 2.12** Two trapezoidal fuzzy numbers $\tilde{A}_1 = (a_1, b_1, c_1, d_1)$ and $\tilde{A}_2 = (a_2, b_2, c_2, d_2)$ are said to be equal i.e., $\tilde{A}_1 = \tilde{A}_2$ if and only if $a_1 = a_2, b_1 = b_2, c_1 = c_2$ and $d_1 = d_2$.

## 2.1.2 Arithmetic Operations

In this section, the arithmetic operations for trapezoidal fuzzy numbers and intervals are presented.

### 2.1.2.1 Arithmetic Operations for Trapezoidal Fuzzy Numbers

In this section, some arithmetic operations for two trapezoidal fuzzy numbers, defined on universal set of real numbers $\mathbb{R}$, are presented.

(i) Let $\tilde{A}_1 = (a_1, b_1, c_1, d_1)$ and $\tilde{A}_2 = (a_2, b_2, c_2, d_2)$ be two trapezoidal fuzzy numbers. Then, $\tilde{A}_1 \oplus \tilde{A}_2 = (a_1 + a_2, b_1 + b_2, c_1 + c_2, d_1 + d_2)$.

(ii) Let $\tilde{A}_1 = (a_1, b_1, c_1, d_1)$ and $\tilde{A}_2 = (a_2, b_2, c_2, d_2)$ be two trapezoidal fuzzy numbers. Then, $\tilde{A}_1 \ominus \tilde{A}_2 = (a_1 - d_2, b_1 - c_2, c_1 - b_2, d_1 - a_2)$.

(iii) Let $\tilde{A}_1 = (a_1, b_1, c_1, d_1)$ and $\tilde{A}_2 = (a_2, b_2, c_2, d_2)$ be two non-negative trapezoidal fuzzy numbers. Then, $\tilde{A}_1 \otimes \tilde{A}_2 = (a_1 a_2, b_1 b_2, c_1 c_2, d_1 d_2)$.

(iv) Let $\tilde{A} = (a, b, c, d)$ be any trapezoidal fuzzy number. Then,

$$\gamma \tilde{A} = \begin{cases} (\gamma a, \gamma b, \gamma c, \gamma d) & \gamma \geq 0 \\ (\gamma d, \gamma c, \gamma b, \gamma a) & \gamma \leq 0 \end{cases}$$

### 2.1.2.2 Arithmetic Operations for Intervals

In this section, some arithmetic operations for two intervals are presented.

(i) Let $A_1 = [a_1, b_1]$ and $A_2 = [a_2, b_2]$ be two intervals. Then,
$A_1 + A_2 = [a_1 + a_2, b_1 + b_2]$

(ii) Let $A_1 = [a_1, b_1]$ and $A_2 = [a_2, b_2]$ be two non-negative intervals. Then,
$A_1 A_2 = [a_1 a_2, b_1 b_2]$

*Remark 2.1* An interval $A = [a, b]$ is said to be non-negative interval if and only if $a \geq 0$.

*Remark 2.2* If $b = c$ then a trapezoidal fuzzy number $(a, b, c, d)$ is said to be triangular fuzzy number and is denoted as $(a, b, b, d)$ or $(a, c, c, d)$ or $(a, b, d)$ or $(a, c, d)$.

*Remark 2.3* [1] Let $\tilde{a}^\lambda = [\underline{a}(\lambda), \overline{a}(\lambda)]$ be a parametric form of an asymmetric triangular fuzzy number $\tilde{a}$ then its nearest symmetric triangular fuzzy number is $(a_0, \sigma)$, where the core '$a_0$' and the spread '$\sigma$' of the symmetric triangular fuzzy number can be obtained as $a_0 = \frac{3}{2} \int_0^1 (\overline{a}(\lambda) - \underline{a}(\lambda))(1 - \lambda) d\lambda$, $\sigma = \frac{1}{2} \int_0^1 (\overline{a}(\lambda) + \underline{a}(\lambda)) d\lambda$.

*Remark 2.4* In the entire thesis 'minimum' and 'maximum' are represented by 'min' and 'max' respectively.

## 2.2 Existing Method for Solving Fully Fuzzy Linear Programming Problems with Equality Constraints

Lotfi et al. [1] pointed out that there is no method in literature for solving fully fuzzy linear programming problems with equality constraints and proposed the following method to find the fuzzy optimal solution of fully fuzzy linear programming problems with equality constraints (2.1):

$$\text{Maximize} \sum_{j=1}^{n} \tilde{c}_j \otimes \tilde{x}_j$$

subject to

$$\sum_{j=1}^{n} \tilde{a}_{ij} \otimes \tilde{x}_j = \tilde{b}_i \ \forall \, i = 1, 2, \ldots, m \tag{2.1}$$

where $\tilde{c}_j$, $\tilde{x}_j$, $\tilde{a}_{ij}$ and $\tilde{b}_i$ are non-negative triangular fuzzy numbers.

**Step 1** Assuming $\tilde{c}_j = (p_j, q_j, r_j)$, $\tilde{a}_{ij} = (a_{ij}, b_{ij}, c_{ij})$, $\tilde{x}_j = (x_j, y_j, z_j)$ and $\tilde{b}_i = (b_i, g_i, h_i)$ the fully fuzzy linear programming problem (2.1) can be written as:

$$\text{Maximize} \sum_{j=1}^{n} (p_j, q_j, r_j) \otimes (x_j, y_j, z_j)$$

subject to

$$\sum_{j=1}^{n} (a_{ij}, b_{ij}, c_{ij}) \otimes (x_j, y_j, z_j) = (b_i, g_i, h_i) \ \forall \, i = 1, 2, \ldots, m \tag{2.2}$$

where $(x_j, y_j, z_j)$ is a non-negative triangular fuzzy number.

**Step 2** Using Definitions 2.8 and 2.10, the fully fuzzy linear programming problem (2.2) can be converted into problem (2.3):

$$\text{Maximize} \sum_{j=1}^{n} [p_j + (q_j - p_j)\lambda, \, r_j - (r_j - q_j)\lambda][x_j + (y_j - x_j)\lambda, \, z_j - (z_j - y_j)\lambda]$$

subject to

$$\sum_{j=1}^{n} [a_{ij} + (b_{ij} - a_{ij})\lambda, \, c_{ij} - (c_{ij} - b_{ij})\lambda][x_j + (y_j - x_j)\lambda, \, z_j - (z_j - y_j)\lambda] \tag{2.3}$$

$$= [b_i + (g_i - b_i)\lambda, \, h_i - (h_i - g_i)\lambda] \ \forall \, i = 1, 2, \ldots, m$$

$$x_j \geq 0, y_j - x_j \geq 0, z_j - y_j \geq 0 \ \forall \, j = 1, 2, \ldots, n$$

**Step 3** Using the arithmetic operations of intervals, defined in Sect. 2.1.2.2, the problem (2.3) can be converted into the problem (2.4):

$$\text{Maximize } [\sum_{j=1}^{n}(p_j x_j + \lambda(p_j y_j + q_j x_j - 2p_j x_j) + \lambda^2(q_j - p_j)(y_j - x_j)), \sum_{j=1}^{n}(r_j z_j - \lambda(2r_j z_j$$

$$- r_j y_j - q_j z_j) + \lambda^2(r_j - q_j)(z_j - y_j))]$$

subject to

$$\sum_{j=1}^{n}[a_{ij}x_j + \lambda(a_{ij}y_j + b_{ij}x_j - 2a_{ij}x_j) + \lambda^2(b_{ij} - a_{ij})(y_j - x_j), c_{ij}z_j - \lambda(2c_{ij}z_j - c_{ij}y_j$$

$$- b_{ij}z_j) + \lambda^2(c_{ij} - b_{ij})(z_j - y_j)] = [b_i + \lambda(g_i - b_i), h_i - \lambda(h_i - g_i)] \ \forall \, i = 1, 2, \ldots, m$$

$$x_j \geq 0, y_j - x_j \geq 0, z_j - y_j \geq 0 \ \forall \, j = 1, 2, \ldots, n$$

(2.4)

**Step 4** Using Remark 2.3, the problem (2.4) can be converted into the problem (2.5):

$$\text{Maximize } (\sum_{j=1}^{n}(\frac{1}{3}q_j y_j + \frac{1}{12}q_j z_j + \frac{1}{12}r_j y_j + \frac{1}{6}r_j z_j + \frac{1}{12}q_j x_j + \frac{1}{12}p_j y_j + \frac{1}{6}p_j x_j), \sum_{j=1}^{n}(\frac{1}{8}q_j z_j$$

$$+ \frac{1}{8}r_j y_j + \frac{3}{8}r_j z_j - \frac{1}{8}q_j x_j - \frac{1}{8}p_j y_j - \frac{3}{8}p_j x_j))$$

subject to

$$\sum_{j=1}^{n}(\frac{1}{3}b_{ij}y_j + \frac{1}{12}b_{ij}z_j + \frac{1}{12}c_{ij}y_j + \frac{1}{6}c_{ij}z_j + \frac{1}{12}b_{ij}x_j + \frac{1}{12}a_{ij}y_j + \frac{1}{6}a_{ij}x_j, \frac{1}{8}b_{ij}z_j + \frac{1}{8}c_{ij}y_j$$

$$+ \frac{3}{8}c_{ij}z_j - \frac{1}{8}b_{ij}x_j - \frac{1}{8}a_{ij}y_j - \frac{3}{8}a_{ij}x_j) = (\frac{1}{4}b_i + \frac{1}{2}g_i + \frac{1}{4}h_i, \frac{1}{2}h_i - \frac{1}{2}b_i) \ \forall \, i = 1, 2, \ldots, m$$

$$x_j \geq 0, y_j - x_j \geq 0, z_j - y_j \geq 0 \ \forall \, j = 1, 2, \ldots, n$$

(2.5)

**Step 5** Using Definition 2.12, convert the obtained fully fuzzy linear programming (2.5) into the crisp linear programming problem (2.6) to maximize the core:

$$\text{Maximize } \sum_{j=1}^{n}(\frac{1}{3}q_j y_j + \frac{1}{12}q_j z_j + \frac{1}{12}r_j y_j + \frac{1}{6}r_j z_j + \frac{1}{12}q_j x_j + \frac{1}{12}p_j y_j + \frac{1}{6}p_j x_j)$$

subject to

$$\sum_{j=1}^{n}(\frac{1}{3}b_{ij}y_j + \frac{1}{12}b_{ij}z_j + \frac{1}{12}c_{ij}y_j + \frac{1}{6}c_{ij}z_j + \frac{1}{12}b_{ij}x_j + \frac{1}{12}a_{ij}y_j + \frac{1}{6}a_{ij}x_j) = \frac{1}{4}b_i + \frac{1}{2}g_i + \frac{1}{4}h_i$$

$$\forall \, i = 1, 2, \ldots, m$$

$$\sum_{j=1}^{n}(\frac{1}{8}b_{ij}z_j + \frac{1}{8}c_{ij}y_j + \frac{3}{8}c_{ij}z_j - \frac{1}{8}b_{ij}x_j - \frac{1}{8}a_{ij}y_j - \frac{3}{8}a_{ij}x_j) = \frac{1}{2}h_i - \frac{1}{2}b_i \ \forall \, i = 1, 2, \ldots, m$$

$$x_j \geq 0, y_j - x_j \geq 0, z_j - y_j \geq 0 \ \forall \, j = 1, 2, \ldots, n$$

(2.6)

**Step 6** If the crisp linear programming problem (2.6) has a unique optimal solution $x_j^*, y_j^*$ and $z_j^*$ then the fuzzy optimal solution of (2.1) will be $(x_j^*, y_j^*, z_j^*)$. If it has alternative optimal solutions then solve the crisp linear programming problem (2.7) to minimize spread:

$$\text{Minimize} \sum_{j=1}^{n} (\frac{1}{8}q_j z_j + \frac{1}{8}r_j y_j + \frac{3}{8}r_j z_j - \frac{1}{8}q_j x_j - \frac{1}{8}p_j y_j - \frac{3}{8}p_j x_j)$$

subject to

$$\sum_{j=1}^{n}(\frac{1}{3}b_{ij}y_j + \frac{1}{12}b_{ij}z_j + \frac{1}{12}c_{ij}y_j + \frac{1}{6}c_{ij}z_j + \frac{1}{12}b_{ij}x_j + \frac{1}{12}a_{ij}y_j + \frac{1}{6}a_{ij}x_j) = \frac{1}{4}b_i + \frac{1}{2}g_i + \frac{1}{4}h_i$$

$$\forall i = 1, 2, \ldots, m$$

$$\sum_{j=1}^{n}(\frac{1}{8}b_{ij}z_j + \frac{1}{8}c_{ij}y_j + \frac{3}{8}c_{ij}z_j - \frac{1}{8}b_{ij}x_j - \frac{1}{8}a_{ij}y_j - \frac{3}{8}a_{ij}x_j) = \frac{1}{2}h_i - \frac{1}{2}b_i \ \forall i = 1, 2, \ldots, m$$

(2.7)

$$\sum_{j=1}^{n}(\frac{1}{3}q_j y_j + \frac{1}{12}q_j z_j + \frac{1}{12}r_j y_j + \frac{1}{6}r_j z_j + \frac{1}{12}q_j x_j + \frac{1}{12}p_j y_j + \frac{1}{6}p_j x_j) = a^*$$

$$x_j \geq 0, y_j - x_j \geq 0, z_j - y_j \geq 0 \ \forall j = 1, 2, \ldots, n$$

where $a^*$ is the optimal value of the crisp linear programming problem (2.6).

**Step 7** Let $x_j^*, y_j^*$ and $z_j^*$ be the optimal solution of crisp linear programming problem (2.7). Then, the fuzzy optimal solution of fully fuzzy linear programming problem (2.1) is $(x_j^*, y_j^*, z_j^*)$.

## 2.3 Limitations and Shortcoming of the Existing Method

In this section, the limitations and shortcoming of the existing method [1] are pointed out.

### 2.3.1 Limitations of the Existing Method

In this section, the limitations of the existing method [1] are pointed out.

The existing method [1] can be used to find the non-negative fuzzy optimal solution of such fully fuzzy linear programming problems with equality constraints in which all the parameters are represented by non-negative triangular fuzzy numbers. However, the existing method [1] cannot be used to find the non-negative fuzzy optimal solution of the following problems:

(i)  Fully fuzzy linear programming problems with equality constraints in which all the parameters are represented by non-negative trapezoidal fuzzy numbers:

$$\text{Maximize/Minimize} \sum_{j=1}^{n} \tilde{c}_j \otimes \tilde{x}_j$$

$$\text{subject to} \qquad\qquad\qquad\qquad (2.8)$$

$$\sum_{j=1}^{n} \tilde{a}_{ij} \otimes \tilde{x}_j = \tilde{b}_i \quad \forall i = 1, 2, \ldots, m$$

where $\tilde{c}_j, \tilde{a}_{ij}, \tilde{b}_i$ and $\tilde{x}_j$ are non-negative trapezoidal fuzzy numbers.

*Example 2.1*

Maximize $((1, 2, 3, 4) \otimes \tilde{x}_1 \oplus (2, 3, 4, 5) \otimes \tilde{x}_2)$
subject to
$(0, 1, 2, 3) \otimes \tilde{x}_1 \oplus (1, 2, 3, 4) \otimes \tilde{x}_2 = (2, 10, 24, 44)$
$(1, 2, 3, 4) \otimes \tilde{x}_1 \oplus (0, 1, 2, 3) \otimes \tilde{x}_2 = (1, 8, 21, 40)$

where $\tilde{x}_1$ and $\tilde{x}_2$ are non-negative trapezoidal fuzzy numbers.

(ii)  Fully fuzzy linear programming problems with equality constraints in which some or all the coefficients are either represented by unrestricted triangular or trapezoidal fuzzy numbers:

$$\text{Maximize/Minimize} \sum_{j=1}^{n} \tilde{c}_j \otimes \tilde{x}_j$$

$$\text{subject to} \qquad\qquad\qquad\qquad (2.9)$$

$$\sum_{j=1}^{n} \tilde{a}_{ij} \otimes \tilde{x}_j = \tilde{b}_i \quad \forall i = 1, 2, \ldots, m$$

where $\tilde{c}_j, \tilde{a}_{ij}, \tilde{b}_i$ are unrestricted triangular or trapezoidal fuzzy numbers and $\tilde{x}_j$ is a non-negative triangular or trapezoidal fuzzy number.

*Example 2.2*

Maximize $((1, 6, 9, 12) \otimes \tilde{x}_1 \oplus (2, 3, 8, 9) \otimes \tilde{x}_2)$
subject to
$(2, 3, 4, 5) \otimes \tilde{x}_1 \oplus (1, 2, 3, 4) \otimes \tilde{x}_2 = (6, 16, 30, 48)$
$(-1, 1, 2, 3) \otimes \tilde{x}_1 \oplus (1, 3, 4, 6) \otimes \tilde{x}_2 = (0, 17, 30, 54)$

where $\tilde{x}_1$ and $\tilde{x}_2$ are non-negative trapezoidal fuzzy numbers.

### 2.3.2  Shortcoming of the Existing Method

In this section, the shortcoming of the existing method [1] is pointed out.

For solving the fully fuzzy linear programming problems by using the existing method [1] there is a need to approximate all the coefficients into its nearest symmetric fuzzy numbers. Due to this conversion, the obtained solutions are approximate and do not satisfy the constraints exactly e.g., on solving the fully fuzzy linear programming problem, chosen in Example 2.3, by using the existing method [1], the obtained fuzzy optimal solution is $\tilde{x}_1 = (1.45, 1.45, 3.31)$ and $\tilde{x}_2 = (0.88, 4.32, 4.32)$ which does not satisfy the constraints exactly.

*Example 2.3*

$$\text{Maximize } ((0, 1, 4) \otimes \tilde{x}_1 \oplus (2, 4, 5) \otimes \tilde{x}_2)$$
$$\text{subject to}$$
$$(2, 3, 7) \otimes \tilde{x}_1 \oplus (2, 4, 5) \otimes \tilde{x}_2 = (6, 18, 46)$$
$$(0, 2, 4) \otimes \tilde{x}_1 \oplus (3, 5, 8) \otimes \tilde{x}_2 = (6, 19, 52)$$

where $\tilde{x}_1$ and $\tilde{x}_2$ are non-negative triangular fuzzy numbers.

## 2.4  Product of a Non-negative Trapezoidal Fuzzy Number with Unrestricted Trapezoidal Fuzzy Number

For solving the fully fuzzy linear programming problems, there is a need to find the product of fuzzy coefficients and fuzzy variables. Since, in the fully fuzzy linear programming problems, the fuzzy coefficients are known and the product of fuzzy numbers depends upon the nature of fuzzy numbers. So, in this section, on the basis of nature of fuzzy coefficients product, proposed by Kumar et al. [2], with the help of existing product of two fuzzy numbers [3], is presented.

Let $\tilde{A} = (a, b, c, d)$ be an unrestricted trapezoidal fuzzy number and $\tilde{X} = (x, y, z, w)$ be a non-negative trapezoidal fuzzy number. Then

$$\tilde{A} \otimes \tilde{X} = \begin{cases} (ax, by, cz, dw) & a \geq 0 \\ (aw, by, cz, dw) & a < 0 \text{ and } b \geq 0 \\ (aw, bz, cz, dw) & b < 0 \text{ and } c \geq 0 \\ (aw, bz, cy, dw) & c < 0 \text{ and } d \geq 0 \\ (aw, bz, cy, dx) & \text{otherwise.} \end{cases}$$

## 2.5 Kumar et al.'s Method to Find the Non-negative Fuzzy Optimal Solution of Fully Fuzzy Linear Programming Problems with Equality Constraints

In this section, to overcome the limitations as well as to resolve the shortcoming of the existing method [1], discussed in Sect. 2.3, Kumar et al.'s method [2] is presented to find the non-negative fuzzy optimal solution of fully fuzzy linear programming problems with equality constraints (2.9).

The steps of the method are as follows:

**Step 1** Assuming $\tilde{c}_j = (p_j, q_j, r_j, s_j)$, $\tilde{x}_j = (x_j, y_j, z_j, w_j)$, $\tilde{a} = (a_{ij}, b_{ij}, c_{ij}, d_{ij})$ and $\tilde{b} = (b_i, g_i, h_i, k_i)$ the fully fuzzy linear programming problem (2.9) can be converted into (2.10):

$$\text{Maximize/Minimize} \sum_{j=1}^{n} (p_j, q_j, r_j, s_j) \otimes (x_j, y_j, z_j, w_j)$$

subject to                                                                      (2.10)

$$\sum_{j=1}^{n} (a_{ij}, b_{ij}, c_{ij}, d_{ij}) \otimes (x_j, y_j, z_j, w_j) = (b_i, g_i, h_i, k_i) \quad \forall \, i = 1, 2, \ldots, m$$

where $(x_j, y_j, z_j, w_j)$ is a non-negative trapezoidal fuzzy number.

**Step 2** Using the product of trapezoidal fuzzy numbers, presented in Sect. 2.4 and assuming $(a_{ij}, b_{ij}, c_{ij}, d_{ij}) \otimes (x_j, y_j, z_j, w_j) = (a'_{ij}, b'_{ij}, c'_{ij}, d'_{ij})$, the fully fuzzy linear programming problem (2.10) can be converted into (2.11):

$$\text{Maximize/Minimize} \sum_{j=1}^{n} (p_j, q_j, r_j, s_j) \otimes (x_j, y_j, z_j, w_j)$$

subject to                                                                      (2.11)

$$\sum_{j=1}^{n} (a'_{ij}, b'_{ij}, c'_{ij}, d'_{ij}) = (b_i, g_i, h_i, k_i) \quad \forall \, i = 1, 2, \ldots, m$$

where $(x_j, y_j, z_j, w_j)$ is a non-negative trapezoidal fuzzy number.

**Step 3** Using arithmetic operations, defined in Sect. 2.1.2.1 and Definition 2.12, the fully fuzzy linear programming problem (2.11) can be converted into (2.12):

$$\text{Maximize/Minimize} \sum_{j=1}^{n} (p_j, q_j, r_j, s_j) \otimes (x_j, y_j, z_j, w_j)$$

subject to

$$\sum_{j=1}^{n} a'_{ij} = b_i \;\; \forall\, i = 1, 2, \ldots, m$$

$$\sum_{j=1}^{n} b'_{ij} = g_i \;\; \forall\, i = 1, 2, \ldots, m \tag{2.12}$$

$$\sum_{j=1}^{n} c'_{ij} = h_i \;\; \forall\, i = 1, 2, \ldots, m$$

$$\sum_{j=1}^{n} d'_{ij} = k_i \;\; \forall\, i = 1, 2, \ldots, m$$

$$x_j \geq 0,\, y_j - x_j \geq 0,\, z_j - y_j \geq 0,\, w_j - z_j \geq 0 \;\; \forall\, j = 1, 2, \ldots, n$$

**Step 4** Suppose the fuzzy linear programming problem (2.12) have '$l$' basic feasible solutions and $\{x^t_j, y^t_j, z^t_j, w^t_j\}$ is the $t$th basic feasible solution then our aim is to find that basic feasible solution out of all '$l$' basic feasible solutions corresponding to which the value of objective function is maximum (or minimum) i.e., our aim is to find $\max_{1 \leq t \leq l}$ (or $\min$)$\{\sum_{j=1}^{n} (p_j, q_j, r_j, s_j) \otimes (x^t_j, y^t_j, z^t_j, w^t_j)\}$. Liou and Wang [4] proposed the concept that if $\max_{1 \leq t \leq l}$ (or $\min$)$\{\Re(\sum_{j=1}^{n} (p_j, q_j, r_j, s_j) \otimes (x^t_j, y^t_j, z^t_j, w^t_j))\}$ is $\Re(\sum_{j=1}^{n} (p_j, q_j, r_j, s_j) \otimes (x^\theta_j, y^\theta_j, z^\theta_j, w^\theta_j))$ then $\max_{1 \leq t \leq l}$ (or $\min$)$\{\sum_{j=1}^{n} (p_j, q_j, r_j, s_j) \otimes (x^t_j, y^t_j, z^t_j, w^t_j)\}$ will also be $\sum_{j=1}^{n} (p_j, q_j, r_j, s_j) \otimes (x^\theta_j, y^\theta_j, z^\theta_j, w^\theta_j)$, where $\Re(a, b, c, d) = \frac{1}{4}(a + b + c + d)$, i.e., according to the existing method [4], the fuzzy optimal solution of (2.12) can be obtained by solving the crisp linear programming problem (2.13):

$$\text{Maximize/Minimize } \Re(\sum_{j=1}^{n}(p_j, q_j, r_j, s_j) \otimes (x_j, y_j, z_j, w_j))$$

subject to

$$\sum_{j=1}^{n} a'_{ij} = b_i \ \forall \, i = 1, 2, \ldots, m$$

$$\sum_{j=1}^{n} b'_{ij} = g_i \ \forall \, i = 1, 2, \ldots, m \tag{2.13}$$

$$\sum_{j=1}^{n} c'_{ij} = h_i \ \forall \, i = 1, 2, \ldots, m$$

$$\sum_{j=1}^{n} d'_{ij} = k_i \ \forall \, i = 1, 2, \ldots, m$$

$$x_j \geq 0, y_j - x_j \geq 0, z_j - y_j \geq 0, w_j - z_j \geq 0 \ \forall \, j = 1, 2, \ldots, n$$

**Step 5** Assuming $(p_j, q_j, r_j, s_j) \otimes (x_j, y_j, z_j, w_j) = (p'_j, q'_j, r'_j, s'_j)$ the crisp linear programming problem (2.13) can be written as (2.14):

$$\text{Maximize/Minimize } \Re(\sum_{j=1}^{n}(p'_j, q'_j, r'_j, s'_j))$$

subject to

$$\sum_{j=1}^{n} a'_{ij} = b_i \ \forall \, i = 1, 2, \ldots, m$$

$$\sum_{j=1}^{n} b'_{ij} = g_i \ \forall \, i = 1, 2, \ldots, m \tag{2.14}$$

$$\sum_{j=1}^{n} c'_{ij} = h_i \ \forall \, i = 1, 2, \ldots, m$$

$$\sum_{j=1}^{n} d'_{ij} = k_i \ \forall \, i = 1, 2, \ldots, m$$

$$x_j \geq 0, y_j - x_j \geq 0, z_j - y_j \geq 0, w_j - z_j \geq 0 \ \forall \, j = 1, 2, \ldots, n$$

**Step 6** Using the linearity property $\Re(\sum_{j=1}^{n} \tilde{A}_i) = \sum_{j=1}^{n} \Re(\tilde{A}_i)$, where $\tilde{A}_i$ is a fuzzy number, the crisp linear programming problem (2.14) can be converted into (2.15):

$$\text{Maximize/Minimize} \sum_{j=1}^{n} \Re(p'_j, q'_j, r'_j, s'_j)$$

subject to

$$\sum_{j=1}^{n} a'_{ij} = b_i \ \forall\, i = 1, 2, \ldots, m$$

$$\sum_{j=1}^{n} b'_{ij} = g_i \ \forall\, i = 1, 2, \ldots, m \tag{2.15}$$

$$\sum_{j=1}^{n} c'_{ij} = h_i \ \forall\, i = 1, 2, \ldots, m$$

$$\sum_{j=1}^{n} d'_{ij} = k_i \ \forall\, i = 1, 2, \ldots, m$$

$$x_j \geq 0,\, y_j - x_j \geq 0,\, z_j - y_j \geq 0,\, w_j - z_j \geq 0 \ \forall\, j = 1, 2, \ldots, n$$

**Step 7** Using $\Re(a, b, c, d) = \frac{1}{4}(a + b + c + d)$ the crisp linear programming problem (2.15) can be converted into (2.16):

$$\text{Maximize/Minimize} \sum_{j=1}^{n} \frac{1}{4}(p'_j + q'_j + r'_j + s'_j)$$

subject to

$$\sum_{j=1}^{n} a'_{ij} = b_i \ \forall\, i = 1, 2, \ldots, m$$

$$\sum_{j=1}^{n} b'_{ij} = g_i \ \forall\, i = 1, 2, \ldots, m \tag{2.16}$$

$$\sum_{j=1}^{n} c'_{ij} = h_i \ \forall\, i = 1, 2, \ldots, m$$

$$\sum_{j=1}^{n} d'_{ij} = k_i \ \forall\, i = 1, 2, \ldots, m$$

$$x_j \geq 0,\, y_j - x_j \geq 0,\, z_j - y_j \geq 0,\, w_j - z_j \geq 0 \ \forall\, j = 1, 2, \ldots, n$$

**Step 8** Solve the crisp linear programming problem (2.16) by using an appropriate existing method [5] to find the optimal solution $\{x^*_j, y^*_j, z^*_j, w^*_j\}$.

**Step 9** Find the fuzzy optimal solution $\{\tilde{x}^*_j\}$ of the fully fuzzy linear programming problem (2.9) by putting the values of $x^*_j, y^*_j, z^*_j$ and $w^*_j$ in $\tilde{x}^*_j = (x^*_j, y^*_j, z^*_j, w^*_j)$.

**Step 10** Find the fuzzy optimal value by putting the values of $\tilde{x}_j^*$, obtained from Step 9, in $\sum_{j=1}^{n} \tilde{c}_j \otimes \tilde{x}_j^*$.


## 2.6   Illustrative Examples

In this section, Kumar et al.'s method [2], presented in Sect. 2.5, is illustrated with the help of fully fuzzy linear programming problems, chosen in Examples 2.1 and 2.2, which cannot be solved by using the existing method [1]. Moreover, a fully fuzzy linear programming problem, which can be solved by using the existing method [1], is also solved by using the Kumar et al.'s method [2].


### 2.6.1   Fuzzy Optimal Solution of the Chosen Fully Fuzzy Linear Programming Problems

In this section, fully fuzzy linear programming problems, chosen in Examples 2.1, 2.2 and 2.3, are solved by using the method, presented is Sect. 2.5.


#### 2.6.1.1   Fuzzy Optimal Solution of the Fully Fuzzy Linear Programming Problem Chosen in Example 2.1

The fuzzy optimal solution of the fully fuzzy linear programming problem, chosen in Example 2.1, can be obtained by using the following steps:

**Step 1** Assuming $\tilde{x}_1 = (x_1, y_1, z_1, w_1)$ and $\tilde{x}_2 = (x_2, y_2, z_2, w_2)$ the fully fuzzy linear programming problem, chosen in Example 2.1, can be written as:

Maximize $((1, 2, 3, 4) \otimes (x_1, y_1, z_1, w_1) \oplus (2, 3, 4, 5) \otimes (x_2, y_2, z_2, w_2))$

subject to

$(0, 1, 2, 3) \otimes (x_1, y_1, z_1, w_1) \oplus (1, 2, 3, 4) \otimes (x_2, y_2, z_2, w_2) = (2, 10, 24, 44)$

$(1, 2, 3, 4) \otimes (x_1, y_1, z_1, w_1) \oplus (0, 1, 2, 3) \otimes (x_2, y_2, z_2, w_2) = (1, 8, 21, 40)$

where $(x_1, y_1, z_1, w_1)$ and $(x_2, y_2, z_2, w_2)$ are non-negative trapezoidal fuzzy numbers.

**Step 2** Using the product, presented in Sect. 2.4, the fully fuzzy linear programming problem, obtained in Step 1, can be written as:

Maximize $((x_1, 2y_1, 3z_1, 4w_1) \oplus (2x_2, 3y_2, 4z_2, 5w_2))$

subject to

$(0x_1, y_1, 2z_1, 3w_1) \oplus (x_2, 2y_2, 3z_2, 4w_2) = (2, 10, 24, 44)$

$(x_1, 2y_1, 3z_1, 4w_1) \oplus (0x_2, y_2, 2z_2, 3w_2) = (1, 8, 21, 40)$

where $(x_1, y_1, z_1, w_1)$ and $(x_2, y_2, z_2, w_2)$ are non-negative trapezoidal fuzzy numbers.

**Step 3** Using the arithmetic operations defined in Sect. 2.1.2.1 and Definition 2.12, the fully fuzzy linear programming problem, obtained in Step 2, can be written as:

Maximize $(x_1 + 2x_2, 2y_1 + 3y_2, 3z_1 + 4z_2, 4w_1 + 5w_2)$

subject to

$$0x_1 + x_2 = 2$$
$$x_1 + 0x_2 = 1$$
$$y_1 + 2y_2 = 10$$
$$2y_1 + y_2 = 8$$
$$2z_1 + 3z_2 = 24$$
$$3z_1 + 2z_2 = 21$$
$$3w_1 + 4w_2 = 44$$
$$4w_1 + 3w_2 = 40$$
$$x_1 \geq 0, y_1 - x_1 \geq 0, z_1 - y_1 \geq 0, w_1 - z_1 \geq 0$$
$$x_2 \geq 0, y_2 - x_2 \geq 0, z_2 - y_2 \geq 0, w_2 - z_2 \geq 0$$

**Step 4** Using Step 4 of the method, presented in Sect. 2.5, the fuzzy linear programming problem, obtained in Step 3, can be written as:

Maximize $\dfrac{1}{4}(x_1 + 2x_2 + 2y_1 + 3y_2 + 3z_1 + 4z_2 + 4w_1 + 5w_2)$

subject to

$$0x_1 + x_2 = 2$$
$$x_1 + 0x_2 = 1$$
$$y_1 + 2y_2 = 10$$
$$2y_1 + y_2 = 8$$
$$2z_1 + 3z_2 = 24$$
$$3z_1 + 2z_2 = 21$$
$$3w_1 + 4w_2 = 44$$
$$4w_1 + 3w_2 = 40$$
$$x_1 \geq 0, y_1 - x_1 \geq 0, z_1 - y_1 \geq 0, w_1 - z_1 \geq 0$$
$$x_2 \geq 0, y_2 - x_2 \geq 0, z_2 - y_2 \geq 0, w_2 - z_2 \geq 0$$

**Step 5** The optimal solution of the crisp linear programming problem, obtained in Step 4, is $x_1 = 1, y_1 = 2, z_1 = 3, w_1 = 4, x_2 = 2, y_2 = 4, z_2 = 6$ and $w_2 = 8$.

**Step 6** Putting the values of $x_1, y_1, z_1, w_1, x_2, y_2, z_2$ and $w_2$ in $\tilde{x}_1 = (x_1, y_1, z_1, w_1)$ and $\tilde{x}_2 = (x_2, y_2, z_2, w_2)$, the exact fuzzy optimal solution is $\tilde{x}_1 = (1, 2, 3, 4), \tilde{x}_2 = (2, 4, 6, 8)$.

**Step 7** Putting the values of $\tilde{x}_1$ and $\tilde{x}_2$, obtained from Step 6, in the objective function the fuzzy optimal value of the fully fuzzy linear programming problem is $(5, 16, 33, 56)$.

### 2.6.1.2 Fuzzy Optimal Solution of the Fully Fuzzy Linear Programming Problem Chosen in Example 2.2

The fuzzy optimal solution of the fully fuzzy linear programming problem, chosen in Example 2.2, can be obtained by using the following steps:

**Step 1** Assuming $\tilde{x}_1 = (x_1, y_1, z_1, w_1)$ and $\tilde{x}_2 = (x_2, y_2, z_2, w_2)$ the fully fuzzy linear programming problem, chosen in Example 2.2, can be written as:

Maximize $((1, 6, 9, 12) \otimes (x_1, y_1, z_1, w_1) \oplus (2, 3, 8, 9) \otimes (x_2, y_2, z_2, w_2))$

subject to

$(2, 3, 4, 5) \otimes (x_1, y_1, z_1, w_1) \oplus (1, 2, 3, 4) \otimes (x_2, y_2, z_2, w_2) = (6, 16, 30, 48)$

$(-1, 1, 2, 3) \otimes (x_1, y_1, z_1, w_1) \oplus (1, 3, 4, 6) \otimes (x_2, y_2, z_2, w_2) = (0, 17, 30, 54)$

where $(x_1, y_1, z_1, w_1)$ and $(x_2, y_2, z_2, w_2)$ are non-negative trapezoidal fuzzy numbers.

**Step 2** Using the product, presented in Sect. 2.4, the fully fuzzy linear programming problem, obtained in Step 1, can be written as:

Maximize $((x_1, 6y_1, 9z_1, 12w_1) \oplus (2x_2, 3y_2, 8z_2, 9w_2))$

subject to

$(2x_1, 3y_1, 4z_1, 5w_1) \oplus (x_2, 2y_2, 3z_2, 4w_2) = (6, 16, 30, 48)$

$(-w_1, y_1, 2z_1, 3w_1) \oplus (x_2, 3y_2, 4z_2, 6w_2) = (0, 17, 30, 54)$

where $(x_1, y_1, z_1, w_1)$ and $(x_2, y_2, z_2, w_2)$ are non-negative trapezoidal fuzzy numbers.

**Step 3** Using the arithmetic operations, defined in Sect. 2.1.2.1 and Definition 2.12, the fuzzy linear programming problem, obtained in Step 2, can be written as:

Maximize $(x_1 + 2x_2, 6y_1 + 3y_2, 9z_1 + 8z_2, 12w_1 + 9w_2)$

subject to

$$2x_1 + x_2 = 6$$
$$-w_1 + x_2 = 0$$
$$3y_1 + 2y_2 = 16$$
$$y_1 + 3y_2 = 17$$
$$4z_1 + 3z_2 = 30$$
$$2z_1 + 4z_2 = 30$$
$$5w_1 + 4w_2 = 48$$
$$3z_1 + 6z_2 = 54$$
$$x_1 \geq 0, y_1 - x_1 \geq 0, z_1 - y_1 \geq 0, w_1 - z_1 \geq 0$$
$$x_2 \geq 0, y_2 - x_2 \geq 0, z_2 - y_2 \geq 0, w_2 - z_2 \geq 0$$

**Step 4** Using Step 4 of the method, presented in Sect. 2.5, the fuzzy linear programming problem, obtained in Step 3, can be written as:

Maximize $\dfrac{1}{4}(x_1 + 2x_2 + 6y_1 + 3y_2 + 9z_1 + 8z_2 + 12w_1 + 9w_2)$

subject to

$$2x_1 + x_2 = 6$$
$$-w_1 + x_2 = 0$$
$$3y_1 + 2y_2 = 16$$
$$y_1 + 3y_2 = 17$$
$$4z_1 + 3z_2 = 30$$
$$2z_1 + 4z_2 = 30$$
$$5w_1 + 4w_2 = 48$$
$$3z_1 + 6z_2 = 54$$
$$x_1 \geq 0, y_1 - x_1 \geq 0, z_1 - y_1 \geq 0, w_1 - z_1 \geq 0$$
$$x_2 \geq 0, y_2 - x_2 \geq 0, z_2 - y_2 \geq 0, w_2 - z_2 \geq 0$$

**Step 5** The optimal solution of the crisp linear programming problem, obtained in Step 4, is $x_1 = 1, y_1 = 2, z_1 = 3, w_1 = 4, x_2 = 4, y_2 = 5, z_2 = 6$ and $w_2 = 7$.

**Step 6** Putting the values of $x_1, y_1, z_1, w_1, x_2, y_2, z_2$ and $w_2$ in $\tilde{x}_1 = (x_1, y_1, z_1, w_1)$ and $\tilde{x}_2 = (x_2, y_2, z_2, w_2)$, the exact fuzzy optimal solution is $\tilde{x}_1 = (1, 2, 3, 4), \tilde{x}_2 = (4, 5, 6, 7)$.

**Step 7** Putting the values of $\tilde{x}_1$ and $\tilde{x}_2$, obtained from Step 6, in the objective function the fuzzy optimal value of the fully fuzzy linear programming problem is $(9, 27, 75, 111)$.

### 2.6.1.3 Fuzzy Optimal Solution of the Fully Fuzzy Linear Programming Problem Chosen in Example 2.3

The fuzzy optimal solution of the fully fuzzy linear programming problem, chosen in Example 2.3, can be obtained by using the following steps:

**Step 1** Assuming $\tilde{x}_1 = (x_1, y_1, z_1)$ and $\tilde{x}_2 = (x_2, y_2, z_2)$ the fully fuzzy linear programming problem, chosen in Example 2.3, can be written as:

Maximize $((0, 1, 4) \otimes (x_1, y_1, z_1) \oplus (2, 4, 5) \otimes (x_2, y_2, z_2))$

subject to

$(2, 3, 7) \otimes (x_1, y_1, z_1) \oplus (2, 4, 5) \otimes (x_2, y_2, z_2) = (6, 18, 46)$

$(0, 2, 4) \otimes (x_1, y_1, z_1) \oplus (3, 5, 8) \otimes (x_2, y_2, z_2) = (6, 19, 52)$

where $(x_1, y_1, z_1)$ and $(x_2, y_2, z_2)$ are non-negative triangular fuzzy numbers.

**Step 2** Using the product, presented in Sect. 2.4, the fully fuzzy linear programming problem, obtained in Step 1, can be written as:

Maximize $((0x_1, y_1, 4z_1) \oplus (2x_2, 4y_2, 5z_2))$

subject to

$(2x_1, 3y_1, 7z_1) \oplus (2x_2, 4y_2, 5z_2) = (6, 18, 46)$

$(0x_1, 2y_1, 4z_1) \oplus (3x_2, 5y_2, 8z_2) = (6, 19, 52)$

where $(x_1, y_1, z_1)$ and $(x_2, y_2, z_2)$ are non-negative triangular fuzzy numbers.

**Step 3** Using the arithmetic operations, defined in Sect. 2.1.2.1 and Definition 2.12, the fuzzy linear programming problem, obtained in Step 2, can be written as:

Maximize $(0x_1 + 2x_2, y_1 + 4y_2, 4z_1 + 5z_2)$

subject to

$$2x_1 + 2x_2 = 6$$
$$0x_1 + 3x_2 = 6$$
$$3y_1 + 4y_2 = 18$$
$$2y_1 + 5y_2 = 19$$
$$7z_1 + 5z_2 = 46$$
$$4z_1 + 8z_2 = 52$$
$$x_1 \geq 0, y_1 - x_1 \geq 0, z_1 - y_1 \geq 0$$
$$x_2 \geq 0, y_2 - x_2 \geq 0, z_2 - y_2 \geq 0$$

**Step 4** Using Step 4 of the method, presented in Sect. 2.5, the fuzzy linear programming problem, obtained in Step 3, can be written as:

$$\text{Maximize } \frac{1}{4}(0x_1 + 2x_2 + 2y_1 + 8y_2 + 4z_1 + 5z_2)$$

subject to

$$2x_1 + 2x_2 = 6$$
$$0x_1 + 3x_2 = 6$$
$$3y_1 + 4y_2 = 18$$
$$2y_1 + 5y_2 = 19$$
$$7z_1 + 5z_2 = 46$$
$$4z_1 + 8z_2 = 52$$
$$x_1 \geq 0, y_1 - x_1 \geq 0, z_1 - y_1 \geq 0$$
$$x_2 \geq 0, y_2 - x_2 \geq 0, z_2 - y_2 \geq 0$$

**Step 5** The optimal solution of the crisp linear programming problem, obtained in Step 4, is $x_1 = 1, y_1 = 2, z_1 = 3, x_2 = 2, y_2 = 3$ and $z_2 = 5$.

**Step 6** Putting the values of $x_1, y_1, z_1, x_2, y_2$ and $z_2$ in $\tilde{x}_1 = (x_1, y_1, z_1)$ and $\tilde{x}_2 = (x_2, y_2, z_2)$, the exact fuzzy optimal solution is $\tilde{x}_1 = (1, 2, 3), \tilde{x}_2 = (2, 3, 5)$.

**Step 7** Putting the values of $\tilde{x}_1$ and $\tilde{x}_2$, obtained from Step 6, in the objective function the fuzzy optimal value of the fully fuzzy linear programming problem is $(4, 14, 37)$.

## 2.7 Advantages of the Kumar et al.'s Method

In this section, the advantages of Kumar et al.'s method, presented in Sect. 2.5, over the existing method [1] are discussed.

(i) It is easy to apply the Kumar et al.'s method, presented in Sect. 2.5, as compared to the existing method [1].

(ii) The fuzzy optimal solution, obtained by using the existing method [1], does not exactly satisfy the constraints of the fully fuzzy linear programming problems while the fuzzy optimal solution, obtained by using the Kumar et al.'s method, presented in Sect. 2.5, exactly satisfy the constraints of the fully fuzzy linear programming problems.

(iii) The existing method [1] can be used to find the non-negative fuzzy optimal solution of fully fuzzy linear programming problems (2.1) but cannot be used to find the non-negative fuzzy optimal solution of fully fuzzy linear programming problems (2.8) and (2.9). However, the Kumar et al.'s method, presented in Sect. 2.5, can be used to find the non-negative fuzzy optimal solution of all the fully fuzzy linear programming problems (2.1), (2.8) and (2.9).

**Table 2.1** Results of the chosen fully fuzzy linear programming problems

| Example | Fuzzy optimal value | |
|---|---|---|
| | Existing method [1] | Kumar et al.'s method |
| 2.1 | Not applicable | (5, 16, 33, 56) |
| 2.2 | Not applicable | (9, 27, 75, 111) |
| 2.3 | (1.76, 18.75, 34.86) | (4, 14, 37) |

## 2.8 Comparative Study

The results of the chosen fully fuzzy linear programming problems, obtained by using the existing method [1] and Kumar et al.'s method, presented in Sect. 2.5, are shown in Table 2.1.

The results, presented in Table 2.1 can be explained as follows:

(i) In the problems, chosen in Examples 2.1 and 2.2, all the coefficients are not non-negative triangular fuzzy numbers. So, due to the limitations of the existing method [1], discussed in Sect. 2.3.1, none of these problems can be solved by using the existing method [1]. However, in the problem, chosen in Example 2.3, all the coefficients are represented by non-negative triangular fuzzy numbers. So, as discussed in Sect. 2.3.1, it can be solved by using the existing method [1] but due to the shortcoming of the existing method [1], discussed in Sect. 2.3.2, the obtained results are not exact.

(ii) The Kumar et al.'s method, presented in Sect. 2.5, can be used to find the non-negative fuzzy optimal solution of fully fuzzy linear programming problems with unrestricted coefficients. So, all the problems, chosen in Examples 2.1, 2.2 and 2.3, can be solved by using the Kumar et al.'s method, presented in Sect. 2.5. Also, as discussed in Sect. 2.7, the results obtained by using the Kumar et al.'s method, presented in Sect. 2.5, are exact.

## 2.9 Conclusions

On the basis of present study, it can be concluded that it is better to use the Kumar et al.'s method [2] as compared to the existing method [1] for solving fully fuzzy linear programming problems with equality constraints.

# References

1. Lotfi, F.H., Allahviranloo, T., Jondabeha, M.A., Alizadeh, L.: Solving a fully fuzzy linear programming using lexicography method and fuzzy approximate solution. Appl. Math. Modell. **33**, 3151–3156 (2009)
2. Kumar, A., Kaur, J., Singh, P.: A new method for solving fully fuzzy linear programming problems. Appl. Math. Modell. **35**, 817–823 (2011)
3. Kaufmann, A., Gupta, M.M.: Introduction to Fuzzy Arithmetic Theory and Applications. Van Nostrand Reinhold, New York (1985)
4. Liou, T.S., Wang, M.J.: Ranking fuzzy numbers with integral value. Fuzzy Sets Syst. **50**, 247–255 (1992)
5. Taha, H.A.: Operations Research: An Introduction. Prentice-Hall, New Jersey (2003)

# Chapter 3
# Fuzzy Optimal Solution of Fully Fuzzy Linear Programming Problems with Equality Constrains

In this chapter, the limitations of the method, presented in Chap. 2, are pointed out and to overcome these limitations, Kaur and Kumar's method [1] is presented for solving fully fuzzy linear programming problems with equality constraints. To show the application of Kaur and Kumar's method [1] a real life problem, which cannot be solved by using the method, presented in Chap. 2, is solved by using the Kaur and Kumar's method [1].

## 3.1 Limitations of the Previous Presented Method

The method, presented in Chap. 2, can be used to find the exact fuzzy optimal solution of the following type of problems:

(i) Fully fuzzy linear programming problems with equality constraints having non-negative fuzzy coefficients and non-negative fuzzy variables.
(ii) Fully fuzzy linear programming problems with equality constraints having unrestricted fuzzy coefficients and non-negative fuzzy variables.

i.e., the method, presented in Chap. 2, can be used to find the exact fuzzy optimal solution of such fully fuzzy linear programming problems with equality constraints in which all the variables are represented by non-negative fuzzy numbers.

However, the method, presented in Chap. 2, cannot be used for solving the following type of problems:

(iii) Fully fuzzy linear programming problems with equality constraints having non-negative fuzzy coefficients and unrestricted fuzzy variables.

---

The contents of this chapter are published in *Applied Intelligence* 37 (2012) 145–154.

© Springer International Publishing Switzerland 2016
J. Kaur and A. Kumar, *An Introduction to Fuzzy Linear Programming Problems*, Studies in Fuzziness and Soft Computing 340,
DOI 10.1007/978-3-319-31274-3_3

$$\text{Maximize/Minimize} \sum_{j=1}^{n} \tilde{c}_j \otimes \tilde{x}_j$$

subject to                                                                 (3.1)

$$\sum_{j=1}^{n} \tilde{a}_{ij} \otimes \tilde{x}_j = \tilde{b}_i \ \forall \ i = 1, 2, \ldots, m$$

where $\tilde{c}_j$, $\tilde{a}_{ij}$ are non-negative trapezoidal fuzzy numbers and $\tilde{b}_i$, $\tilde{x}_j$ are unrestricted trapezoidal fuzzy numbers.

*Example 3.1*

> Maximize $((1, 2, 3, 4) \otimes \tilde{x}_1 \oplus (2, 4, 6, 8) \otimes \tilde{x}_2)$
>
> subject to
>
> $(0, 1, 2, 3) \otimes \tilde{x}_1 \oplus (1, 3, 5, 7) \otimes \tilde{x}_2 = (-8, 2, 27, 57)$
>
> $(2, 4, 7, 9) \otimes \tilde{x}_1 \oplus (2, 3, 5, 6) \otimes \tilde{x}_2 = (-25, -8, 34, 81)$

where $\tilde{x}_1$ and $\tilde{x}_2$ are unrestricted trapezoidal fuzzy numbers.

(iv) Fully fuzzy linear programming problems with equality constraints having unrestricted fuzzy coefficients and unrestricted fuzzy variables:

$$\text{Maximize/Minimize} \sum_{j=1}^{n} \tilde{c}_j \otimes \tilde{x}_j$$

subject to                                                                 (3.2)

$$\sum_{j=1}^{n} \tilde{a}_{ij} \otimes \tilde{x}_j = \tilde{b}_i \ \forall \ i = 1, 2, \ldots, m$$

where $\tilde{c}_j$, $\tilde{a}_{ij}$, $\tilde{b}_i$ and $\tilde{x}_j$ are unrestricted trapezoidal fuzzy numbers.

*Example 3.2*

> Maximize $((-2, -1, 3, 6) \otimes \tilde{x}_1 \oplus (2, 4, 6, 8) \otimes \tilde{x}_2)$
>
> subject to
>
> $(-3, 2, 5, 7) \otimes \tilde{x}_1 \oplus (-2, -1, 5, 7) \otimes \tilde{x}_2 = (-42, -17, 45, 91)$
>
> $(1, 2, 3, 4) \otimes \tilde{x}_1 \oplus (-4, -3, -2, -1) \otimes \tilde{x}_2 = (-48, -27, -4, 28)$

where $\tilde{x}_1$ and $\tilde{x}_2$ are unrestricted trapezoidal fuzzy numbers.

i.e., the method, presented in Chap. 2, cannot be used for solving such fully fuzzy linear programming problems with equality constraints in which all or some of the variables are represented by unrestricted fuzzy numbers.

## 3.2 Product of Unrestricted Trapezoidal Fuzzy Numbers

In this section, the product of two unrestricted trapezoidal fuzzy numbers is presented [1].

Let $\tilde{A} = (a, b, c, d)$ and $\tilde{X} = (x, y, z, w)$ be two unrestricted trapezoidal fuzzy numbers. Then,

$$\tilde{A} \otimes \tilde{X} = \begin{cases} (\min\{ax, dx\}, \min\{by, cy\}, \max\{bz, cz\}, \max\{aw, dw\}) \ a \geq 0 \\ (\min\{aw, dx\}, \min\{by, cy\}, \max\{bz, cz\}, \max\{ax, dw\}) \ a < 0 \text{ and } b \geq 0 \\ (\min\{aw, dx\}, \min\{bz, cy\}, \max\{by, cz\}, \max\{ax, dw\}) \ b < 0 \text{ and } c \geq 0 \\ (\min\{aw, dx\}, \min\{bz, cz\}, \max\{by, cy\}, \max\{ax, dw\}) \ c < 0 \text{ and } d \geq 0 \\ (\min\{aw, dw\}, \min\{bz, cz\}, \max\{by, cy\}, \max\{ax, dx\}) \text{ otherwise.} \end{cases}$$

### 3.2.1 Particular Cases of the Product of Unrestricted Trapezoidal Fuzzy Numbers

In this section, all the particular cases of the product, presented in Sect. 3.2, are discussed.

(i) Let $\tilde{A} = (a, b, c, d)$ be an unrestricted trapezoidal fuzzy number and $\tilde{X} = (x, y, z, w)$ be a non-negative trapezoidal fuzzy number. Then,

$$\tilde{A} \otimes \tilde{X} = \begin{cases} (ax, by, cz, dw) \ a \geq 0 \\ (aw, by, cz, dw) \ a < 0 \text{ and } b \geq 0 \\ (aw, bz, cz, dw) \ b < 0 \text{ and } c \geq 0 \\ (aw, bz, cy, dw) \ c < 0 \text{ and } d \geq 0 \\ (aw, bz, cy, dx) \text{ otherwise.} \end{cases}$$

(ii) Let $\tilde{A} = (a, b, c, d)$ be a non-negative trapezoidal fuzzy number and $\tilde{X} = (x, y, z, w)$ be an unrestricted trapezoidal fuzzy number. Then,

$$\tilde{A} \otimes \tilde{X} = (\min\{ax, dx\}, \min\{by, cy\}, \max\{bz, cz\}, \max\{aw, dw\})$$

(iii) Let $\tilde{A} = (a, b, c, d)$ and $\tilde{X} = (x, y, z, w)$ be two non-negative trapezoidal fuzzy number. Then,

$$\tilde{A} \otimes \tilde{X} = (ax, by, cz, dw)$$

## 3.3   Kaur and Kumar's Method for Solving Fully Fuzzy Linear Programming Problems with Equality Constraints

In this section, to overcome all the limitations of the method, presented in Chap. 2, Kaur and Kumar's method [1] is presented to find the exact fuzzy optimal solution of fully fuzzy linear programming problems (3.2) in which all the parameters are represented by unrestricted trapezoidal fuzzy numbers.

The steps of the method are as follows:

**Step 1** Assuming $\tilde{c}_j = (p_j, q_j, r_j, s_j), \tilde{x}_j = (x_j, y_j, z_j, w_j), \tilde{a}_{ij} = (a_{ij}, b_{ij}, c_{ij}, d_{ij})$ and $\tilde{b}_i = (b_i, g_i, h_i, k_i)$ the fully fuzzy linear programming problem (3.2) can be written as:

$$\text{Maximize/Minimize} \sum_{j=1}^{n} (p_j, q_j, r_j, s_j) \otimes (x_j, y_j, z_j, w_j)$$

subject to

$$\sum_{j=1}^{n} (a_{ij}, b_{ij}, c_{ij}, d_{ij}) \otimes (x_j, y_j, z_j, w_j) = (b_i, g_i, h_i, k_i) \quad \forall\, i = 1, 2, \ldots, m$$

(3.3)

where $(x_j, y_j, z_j, w_j)$ is an unrestricted trapezoidal fuzzy number.

**Step 2** Using the product of trapezoidal fuzzy numbers, presented in Sect. 3.2, the fully fuzzy linear programming problem (3.3) can be written as:

$$\text{Maximize/Minimize} \sum_{j=1}^{n} (p_j, q_j, r_j, s_j) \otimes (x_j, y_j, z_j, w_j)$$

subject to

$$\sum_{j=1}^{n} (\min\{a'_{ij}, a''_{ij}\}, \min\{b'_{ij}, b''_{ij}\}, \max\{c'_{ij}, c''_{ij}\}, \max\{d'_{ij}, d''_{ij}\})$$

$$= (b_i, g_i, h_i, k_i) \qquad \forall\, i = 1, 2, \ldots, m$$

(3.4)

where $(x_j, y_j, z_j, w_j)$ is an unrestricted trapezoidal fuzzy number and

$(\min\{a'_{ij}, a''_{ij}\}, \min\{b'_{ij}, b''_{ij}\}, \max\{c'_{ij}, c''_{ij}\}, \max\{d'_{ij}, d''_{ij}\})$

$$= \begin{cases} \min\{a_{ij}x_j, d_{ij}x_j\}, \min\{b_{ij}y_j, c_{ij}y_j\}, \max\{b_{ij}z_j, c_{ij}z_j\}, \max\{a_{ij}w_j, d_{ij}w_j\}) & a_{ij} \geq 0 \\ \min\{a_{ij}w_j, d_{ij}x_j\}, \min\{b_{ij}y_j, c_{ij}y_j\}, \max\{b_{ij}z_j, c_{ij}z_j\}, \max\{d_{ij}w_j, a_{ij}x_j\}) & a_{ij} < 0 \text{ and } b_{ij} \geq 0 \\ (\min\{a_{ij}w_j, d_{ij}x_j\}, \min\{b_{ij}z_j, c_{ij}y_j\}, \max\{b_{ij}y_j, c_{ij}z_j\}, \min\{a_{ij}x_j, d_{ij}w_j\}) & b_{ij} < 0 \text{ and } c_{ij} \geq 0 \\ (\min\{a_{ij}w_j, d_{ij}x_j\}, \min\{b_{ij}z_j, c_{ij}z_j\}, \max\{b_{ij}y_j, c_{ij}y_j\}, \max\{a_{ij}x_j, d_{ij}w_j\}) & c_{ij} < 0 \text{ and } d_{ij} \geq 0 \\ (\min\{a_{ij}w_j, d_{ij}w_j\}, \min\{b_{ij}z_j, c_{ij}z_j\}, \max\{b_{ij}y_j, c_{ij}y_j\}, \max\{a_{ij}x_j, d_{ij}x_j\}) & \text{otherwise.} \end{cases}$$

**Step 3** Using arithmetic operations, defined in Sect. 2.1.2.1 and Definition 2.12, the fully fuzzy linear programming problem (3.4) can be written as:

$$\text{Maximize/Minimize } \sum_{j=1}^{n} (p_j, q_j, r_j, s_j) \otimes (x_j, y_j, z_j, w_j)$$

subject to

$$\sum_{j=1}^{n} \min\{a'_{ij}, a''_{ij}\} = b_i \quad \forall i = 1, 2, \ldots, m$$

$$\sum_{j=1}^{n} \min\{b'_{ij}, b''_{ij}\} = g_i \quad \forall i = 1, 2, \ldots, m \tag{3.5}$$

$$\sum_{j=1}^{n} \max\{c'_{ij}, c''_{ij}\} = h_i \quad \forall i = 1, 2, \ldots, m$$

$$\sum_{j=1}^{n} \max\{d'_{ij}, d''_{ij}\} = k_i \quad \forall i = 1, 2, \ldots, m$$

$$y_j - x_j \geq 0, z_j - y_j \geq 0, w_j - z_j \geq 0 \ \forall j = 1, 2, \ldots, n$$

**Step 4** As discussed in Step 4 of Sect. 2.5, the fuzzy optimal solution of (3.5) can be obtained by solving the crisp non-linear programming problem (3.6):

$$\text{Maximize/Minimize } \Re\left(\sum_{j=1}^{n} (p_j, q_j, r_j, s_j) \otimes (x_j, y_j, z_j, w_j)\right)$$

subject to

$$\sum_{j=1}^{n} \min\{a'_{ij}, a''_{ij}\} = b_i \quad \forall i = 1, 2, \ldots, m$$

$$\sum_{j=1}^{n} \min\{b'_{ij}, b''_{ij}\} = g_i \quad \forall i = 1, 2, \ldots, m \tag{3.6}$$

$$\sum_{j=1}^{n} \max\{c'_{ij}, c''_{ij}\} = h_i \quad \forall i = 1, 2, \ldots, m$$

$$\sum_{j=1}^{n} \max\{d'_{ij}, d''_{ij}\} = k_i \quad \forall i = 1, 2, \ldots, m$$

$$y_j - x_j \geq 0, z_j - y_j \geq 0, w_j - z_j \geq 0 \ \forall j = 1, 2, \ldots, n$$

**Step 5** Assuming $(p_j, q_j, r_j, s_j) \otimes (x_j, y_j, z_j, w_j) = (\min\{p'_j, p''_j\}, \min\{q'_j, q''_j\},$
$\max\{r'_j, r''_j\}, \max\{s'_j, s''_j\})$ the crisp non-linear programming problem (3.6) can be written as:

$$\text{Maximize/Minimize } \Re(\sum_{j=1}^{n}(\min\{p'_j, p''_j\}, \min\{q'_j, q''_j\}, \max\{r'_j, r''_j\}, \max\{s'_j, s''_j\}))$$

subject to

$$\sum_{j=1}^{n} \min\{a'_{ij}, a''_{ij}\} = b_i \quad \forall i = 1, 2, \ldots, m$$

$$\sum_{j=1}^{n} \min\{b'_{ij}, b''_{ij}\} = g_i \quad \forall i = 1, 2, \ldots, m \tag{3.7}$$

$$\sum_{j=1}^{n} \max\{c'_{ij}, c''_{ij}\} = h_i \quad \forall i = 1, 2, \ldots, m$$

$$\sum_{j=1}^{n} \max\{d'_{ij}, d''_{ij}\} = k_i \quad \forall i = 1, 2, \ldots, m$$

$$y_j - x_j \geq 0, z_j - y_j \geq 0, w_j - z_j \geq 0 \ \forall j = 1, 2, \ldots, n$$

**Step 6** Using the linearity property $\Re(\sum_{j=1}^{n} \tilde{A}_i) = \sum_{j=1}^{n} \Re(\tilde{A}_i)$, where $\tilde{A}_i$ is a fuzzy number, problem (3.7) can be converted into (3.8):

$$\text{Maximize/Minimize } (\sum_{j=1}^{n} \Re(\min\{p'_j, p''_j\}, \min\{q'_j, q''_j\}, \max\{r'_j, r''_j\}, \max\{s'_j, s''_j\}))$$

subject to

$$\sum_{j=1}^{n} \min\{a'_{ij}, a''_{ij}\} = b_i \quad \forall i = 1, 2, \ldots, m$$

$$\sum_{j=1}^{n} \min\{b'_{ij}, b''_{ij}\} = g_i \quad \forall i = 1, 2, \ldots, m \tag{3.8}$$

$$\sum_{j=1}^{n} \max\{c'_{ij}, c''_{ij}\} = h_i \quad \forall i = 1, 2, \ldots, m$$

$$\sum_{j=1}^{n} \max\{d'_{ij}, d''_{ij}\} = k_i \quad \forall i = 1, 2, \ldots, m$$

$$y_j - x_j \geq 0, z_j - y_j \geq 0, w_j - z_j \geq 0 \ \forall j = 1, 2, \ldots, n$$

**Step 7** Using $\max(a, b) = \frac{a+b}{2} + |\frac{a-b}{2}|$ and $\min(a, b) = \frac{a+b}{2} - |\frac{a-b}{2}|$ problem (3.8) can be converted into (3.9):

$$\text{Maximize/Minimize} \sum_{j=1}^{n} \Re(\frac{p'_j + p''_j}{2} - |\frac{p'_j - p''_j}{2}|, \frac{q'_j + q''_j}{2} - |\frac{q'_j - q''_j}{2}|,$$

$$\frac{r'_j + r''_j}{2} + |\frac{r'_j - r''_j}{2}|, \frac{s'_j + s''_j}{2} + |\frac{s'_j - s''_j}{2}|)$$

subject to

$$\sum_{j=1}^{n}(\frac{a'_{ij} + a''_{ij}}{2} - |\frac{a'_{ij} - a''_{ij}}{2}|) = b_i \ \forall i = 1, 2, \ldots, m$$

$$\sum_{j=1}^{n}(\frac{b'_{ij} + b''_{ij}}{2} - |\frac{b'_{ij} - b''_{ij}}{2}|) = g_i \ \forall i = 1, 2, \ldots, m \tag{3.9}$$

$$\sum_{j=1}^{n}(\frac{c'_{ij} + c''_{ij}}{2} + |\frac{c'_{ij} - c''_{ij}}{2}|) = h_i \ \forall i = 1, 2, \ldots, m$$

$$\sum_{j=1}^{n}(\frac{d'_{ij} + d''_{ij}}{2} + |\frac{d'_{ij} - d''_{ij}}{2}|) = k_i \ \forall i = 1, 2, \ldots, m$$

$$y_j - x_j \geq 0, z_j - y_j \geq 0, w_j - z_j \geq 0 \ \forall j = 1, 2, \ldots, n$$

**Step 8** Using $\Re(a, b, c, d) = \frac{1}{4}(a + b + c + d)$ problem (3.9) can be converted into (3.10):

$$\text{Maximize/Minimize} \frac{1}{4}(\sum_{j=1}^{n}(\frac{p'_j + p''_j}{2} - |\frac{p'_j - p''_j}{2}| + \frac{q'_j + q''_j}{2} - |\frac{q'_j - q''_j}{2}| +$$

$$\frac{r'_j + r''_j}{2} + |\frac{r'_j - r''_j}{2}| + \frac{s'_j + s''_j}{2} + |\frac{s'_j - s''_j}{2}|))$$

subject to

$$\sum_{j=1}^{n}(\frac{a'_{ij} + a''_{ij}}{2} - |\frac{a'_{ij} - a''_{ij}}{2}|) = b_i \ \forall i = 1, 2, \ldots, m$$

$$\sum_{j=1}^{n}(\frac{b'_{ij} + b''_{ij}}{2} - |\frac{b'_{ij} - b''_{ij}}{2}|) = g_i \ \forall i = 1, 2, \ldots, m \tag{3.10}$$

$$\sum_{j=1}^{n}(\frac{c'_{ij} + c''_{ij}}{2} + |\frac{c'_{ij} - c''_{ij}}{2}|) = h_i \ \forall i = 1, 2, \ldots, m$$

$$\sum_{j=1}^{n}(\frac{d'_{ij} + d''_{ij}}{2} + |\frac{d'_{ij} - d''_{ij}}{2}|) = k_i \ \forall i = 1, 2, \ldots, m$$

$$y_j - x_j \geq 0, z_j - y_j \geq 0, w_j - z_j \geq 0 \ \forall j = 1, 2, \ldots, n$$

**Step 9** Solve the crisp non-linear programming problem (3.10) by using an appropriate existing method [2] to find the optimal solution $\{x_j^*, y_j^*, z_j^*, w_j^*\}$.

**Step 10** Find the fuzzy optimal solution $\{\tilde{x}_j^*\}$ of the fully fuzzy linear programming problem (3.2) by putting the values of $x_j^*, y_j^*, z_j^*$ and $w_j^*$ in $\tilde{x}_j^* = (x_j^*, y_j^*, z_j^*, w_j^*)$.

**Step 11** Find the fuzzy optimal value by putting the values of $\tilde{x}_j^*$, obtained from Step 10, in $\sum_{j=1}^{n} \tilde{c}_j \otimes \tilde{x}_j^*$.

## 3.4 Illustrative Examples

In this section, the method, presented in Sect. 3.3, is illustrated with the help of fully fuzzy linear programming problems, chosen in Examples 3.1 and 3.2, which cannot be solved by using the method, presented in Chap. 2. Moreover, the fully fuzzy linear programming problem, chosen in Example 2.1, which can be solved by using the method, presented in Chap. 2, is also solved by using the method, presented in Sect. 3.3.

### 3.4.1 Fuzzy Optimal Solution of the Chosen Fully Fuzzy Linear Programming Problems

In this section, fully fuzzy linear programming problems, chosen in Examples 2.1, 3.1 and 3.2, are solved by using the method, presented in Sect. 3.3.

#### 3.4.1.1 Fuzzy Optimal Solution of the Fully Fuzzy Linear Programming Problem Chosen in Example 3.1

The fuzzy optimal solution of the fully fuzzy linear programming problem, chosen in Example 3.1, can be obtained by using the following steps:

**Step 1** Assuming $\tilde{x}_1 = (x_1, y_1, z_1, w_1)$ and $\tilde{x}_2 = (x_2, y_2, z_2, w_2)$ the fully fuzzy linear programming problem, chosen in Example 3.1, can be written as:

Maximize $((1, 2, 3, 4) \otimes (x_1, y_1, z_1, w_1) \oplus (2, 4, 6, 8) \otimes (x_2, y_2, z_2, w_2))$

subject to

$(0, 1, 2, 3) \otimes (x_1, y_1, z_1, w_1) \oplus (1, 3, 5, 7) \otimes (x_2, y_2, z_2, w_2) = (-8, 2, 27, 57)$

$(2, 4, 7, 9) \otimes (x_1, y_1, z_1, w_1) \oplus (2, 3, 5, 6) \otimes (x_2, y_2, z_2, w_2) = (-25, -8, 34, 81)$

where $(x_1, y_1, z_1, w_1)$ and $(x_2, y_2, z_2, w_2)$ are unrestricted fuzzy numbers.

**Step 2** Using the product, presented in Sect. 3.2, the fully fuzzy linear programming problem, obtained in Step 1, can be written as:

Maximize $((\min\{x_1, 4x_1\}, \min\{2y_1, 3y_1\}, \max\{2z_1, 3z_1\}, \max\{w_1, 4w_1\}) \oplus (\min\{2x_2, 8x_2\}, \min\{4y_2, 6y_2\}, \max\{4z_2, 6z_2\}, \max\{2w_2, 8w_2\}))$

subject to

$(\min\{0, 3x_1\}, \min\{y_1, 2y_1\}, \max\{z_1, 2z_1\}, \max\{0, 3w_1\}) \oplus (\min\{x_2, 7x_2\}, \min\{3y_2, 5y_2\}, \max\{3z_2, 5z_2\}, \max\{w_2, 7w_2\}) = (-8, 2, 27, 57)$

$(\min\{2x_1, 9x_1\}, \min\{4y_1, 7y_1\}, \max\{4z_1, 7z_1\}, \max\{2w_1, 9w_1\}) \oplus (\min\{2x_2, 6x_2\}, \min\{3y_2, 5y_2\}\max\{3z_2, 5z_2\}, \max\{2w_2, 6w_2\}) = (-25, -8, 34, 81)$

$$y_1 - x_1 \geq 0, z_1 - y_1 \geq 0, w_1 - z_1 \geq 0$$
$$y_2 - x_2 \geq 0, z_2 - y_2 \geq 0, w_2 - z_2 \geq 0$$

**Step 3** Using the arithmetic operations, defined in Sect. 2.1.2.1 and Definition 2.12, the fully fuzzy linear programming problem, obtained in Step 2, can be written as:

Maximize $(\min\{x_1, 4x_1\} + \min\{2x_2, 8x_2\}, \min\{2y_1, 3y_1\} + \min\{4y_2, 6y_2\}, \max\{2z_1, 3z_1\} + \max\{4z_2, 6z_2\}, \max\{w_1, 4w_1\} + \max\{2w_2, 8w_2\})$

subject to

$$\min\{0, 3x_1\} + \min\{x_2, 7x_2\} = -8$$
$$\min\{y_1, 2y_1\} + \min\{3y_2, 5y_2\} = 2$$
$$\max\{z_1, 2z_1\} + \max\{3z_2, 5z_2\} = 27$$
$$\max\{0, 3w_1\} + \max\{w_2, 7w_2\} = 57$$
$$\min\{2x_1, 9x_1\} + \min\{2x_2, 6x_2\} = -25$$
$$\min\{4y_1, 7y_1\} + \min\{3y_2, 5y_2\} = -8$$
$$\max\{4z_1, 7z_1\} + \max\{3z_2, 5z_2\} = 34$$
$$\max\{2w_1, 9w_1\} + \max\{2w_2, 6w_2\} = 81$$
$$y_1 - x_1 \geq 0, z_1 - y_1 \geq 0, w_1 - z_1 \geq 0$$
$$y_2 - x_2 \geq 0, z_2 - y_2 \geq 0, w_2 - z_2 \geq 0$$

**Step 4** Using Step 4 of the method, presented in Sect. 3.3, the fuzzy linear programming problem, obtained in Step 3, can be written as:

Maximize $\Re(\min\{x_1, 4x_1\} + \min\{2x_2, 8x_2\}, \min\{2y_1, 3y_1\} + \min\{4y_2, 6y_2\}, \max\{2z_1, 3z_1\} + \max\{4z_2, 6z_2\}, \max\{w_1, 4w_1\} + \max\{2w_2, 8w_2\})$

subject to

$$\min\{0, 3x_1\} + \min\{x_2, 7x_2\} = -8$$
$$\min\{y_1, 2y_1\} + \min\{3y_2, 5y_2\} = 2$$
$$\max\{z_1, 2z_1\} + \max\{3z_2, 5z_2\} = 27$$

$$\max\{0, 3w_1\} + \max\{w_2, 7w_2\} = 57$$
$$\min\{2x_1, 9x_1\} + \min\{2x_2, 6x_2\} = -25$$
$$\min\{4y_1, 7y_1\} + \min\{3y_2, 5y_2\} = -8$$
$$\max\{4z_1, 7z_1\} + \max\{3z_2, 5z_2\} = 34$$
$$\max\{2w_1, 9w_1\} + \max\{2w_2, 6w_2\} = 81$$
$$y_1 - x_1 \geq 0, z_1 - y_1 \geq 0, w_1 - z_1 \geq 0$$
$$y_2 - x_2 \geq 0, z_2 - y_2 \geq 0, w_2 - z_2 \geq 0$$

**Step 5** Using Steps 7 and 8 of the method, presented in Sect. 3.3, the problem, obtained in Step 4, can be converted into the following crisp non-linear programming problem:

$$\text{Maximize } \frac{1}{4}(\frac{5}{2}x_1 - \frac{3}{2}|x_1| + 5x_2 - 3|x_2| + \frac{5}{2}y_1 - \frac{1}{2}|y_1| + 5y_2 - |y_2| + \frac{5}{2}z_1 +$$
$$\frac{1}{2}|z_1| + 5z_2 + |z_2| + \frac{5}{2}w_1 + \frac{3}{2}|w_1| + 5w_2 + 3|w_2|)$$

subject to

$$\frac{3}{2}x_1 - \frac{3}{2}|x_1| + 4x_2 - 3|x_2| = -8$$

$$\frac{3}{2}y_1 - \frac{1}{2}|y_1| + 4y_2 - |y_2| = 2$$

$$\frac{3}{2}z_1 + \frac{1}{2}|z_1| + 4z_2 + |z_2| = 27$$

$$\frac{3}{2}w_1 + \frac{3}{2}|w_1| + 4w_2 + 3|w_2| = 57$$

$$\frac{11}{2}x_1 - \frac{7}{2}|x_1| + 4x_2 - 2|x_2| = -25$$

$$\frac{11}{2}y_1 - \frac{3}{2}|y_1| + 4y_2 - |y_2| = -8$$

$$\frac{11}{2}z_1 + \frac{3}{2}|z_1| + 4z_2 + |z_2| = 34$$

$$\frac{11}{2}w_1 + \frac{7}{2}|w_1| + 4w_2 + 2|w_2| = 81$$

$$y_1 - x_1 \geq 0, z_1 - y_1 \geq 0, w_1 - z_1 \geq 0$$
$$y_2 - x_2 \geq 0, z_2 - y_2 \geq 0, w_2 - z_2 \geq 0$$

**Step 6** The optimal solution of the crisp non-linear programming problem, obtained in Step 5, is $x_1 = -3$, $y_1 = -2$, $z_1 = \frac{7}{5}$, $w_1 = 5$, $x_2 = 1$, $y_2 = 2$, $z_2 = \frac{121}{25}$ and $w_2 = 6$.

**Step 7** Putting the values of $x_1, y_1, z_1, w_1, x_2, y_2, z_2$ and $w_2$ in $\tilde{x}_1 = (x_1, y_1, z_1, w_1)$ and $\tilde{x}_2 = (x_2, y_2, z_2, w_2)$, the exact fuzzy optimal solution is $\tilde{x}_1 = (-3, -2, \frac{7}{5}, 5)$, $\tilde{x}_2 = (1, 2, \frac{121}{25}, 6)$.

**Step 8** Putting the values of $\tilde{x}_1$ and $\tilde{x}_2$, obtained from Step 7, in the objective function the fuzzy optimal value is $(-10, 2, \frac{831}{25}, 68)$.

#### 3.4.1.2 Fuzzy Optimal Solution of the Fully Fuzzy Linear Programming Problem Chosen in Example 3.2

The fuzzy optimal solution of the fully fuzzy linear programming problem, chosen in Example 3.2, can be obtained by using the following steps:

**Step 1** Assuming $\tilde{x}_1 = (x_1, y_1, z_1, w_1)$ and $\tilde{x}_2 = (x_2, y_2, z_2, w_2)$ the fully fuzzy linear programming problem, chosen in Example 3.2, can be written as:

Maximize $((-2, -1, 3, 6) \otimes (x_1, y_1, z_1, w_1) \oplus (2, 4, 6, 8) \otimes (x_2, y_2, z_2, w_2))$

subject to

$(-3, 2, 5, 7) \otimes (x_1, y_1, z_1, w_1) \oplus (-2, -1, 5, 7) \otimes (x_2, y_2, z_2, w_2) = (-42, -17, 45, 91)$
$(1, 2, 3, 4) \otimes (x_1, y_1, z_1, w_1) \oplus (-4, -3, -2, -1) \otimes (x_2, y_2, z_2, w_2) = (-48, -27, -4, 28)$

where $(x_1, y_1, z_1, w_1)$ and $(x_2, y_2, z_2, w_2)$ are unrestricted trapezoidal fuzzy numbers.

**Step 2** Using the product, presented in Sect. 3.2, the fully fuzzy linear programming problem, obtained in Step 1, can be written as:

Maximize $((\min\{-2w_1, 6x_1\}, \min\{-z_1, 3y_1\}, \max\{-y_1, 3z_1\}, \max\{-2x_1, 6w_1\}) \oplus (\min\{2x_2, 8x_2\}, \min\{4y_2, 6y_2\}, \max\{4z_2, 6z_2\}, \max\{2w_2, 8w_2\}))$

subject to

$(\min\{-3w_1, 7x_1\}, \min\{2y_1, 5y_1\}, \max\{2z_1, 5z_1\}, \max\{7w_1, -3x_1\}) \oplus (\min\{-2w_2, 7x_2\}, \min\{-z_2, 5y_2\}, \max\{-y_2, 5z_2\}, \max\{-2x_2, 7w_2\}) = (-42, -17, 45, 91)$
$(\min\{x_1, 4x_1\}, \min\{2y_1, 3y_1\}, \max\{2z_1, 3z_1\}, \max\{w_1, 4w_1\}) \oplus (\min\{-4w_2, -w_2\}, \min\{-3z_2, -2z_2\}, \max\{-3y_2, -2y_2\}, \max\{-4x_2, -x_2\}) = (-48, -27, -4, 28)$

$$y_1 - x_1 \geq 0, z_1 - y_1 \geq 0, w_1 - z_1 \geq 0$$
$$y_2 - x_2 \geq 0, z_2 - y_2 \geq 0, w_2 - z_2 \geq 0$$

**Step 3** Using the arithmetic operations, defined in Sect. 2.1.2.1 and Definition 2.12, the fully fuzzy linear programming problem, obtained in Step 2, can be written as:

Maximize $(\min\{-2w_1, 6x_1\} + \min\{2x_2, 8x_2\}, \min\{-z_1, 3y_1\} + \min\{4y_2, 6y_2\}, \max\{-y_1, 3z_1\} + \max\{4z_2, 6z_2\}, \max\{-2x_1, 6w_1\} + \max\{2w_2, 8w_2\})$

subject to

$$\min\{-3w_1, 7x_1\} + \min\{-2w_2, 7x_2\} = -42$$
$$\min\{2y_1, 5y_1\} + \min\{-z_2, 5y_2\} = -17$$
$$\max\{2z_1, 5z_1\} + \max\{-y_2, 5z_2\} = 45$$

$$\max\{7w_1, -3x_1\} + \max\{-2x_2, 7w_2\} = 91$$
$$\min\{x_1, 4x_1\} + \min\{-4w_2, -w_2\} = -48$$
$$\min\{2y_1, 3y_1\} + \min\{-3z_2, -2z_2\} = -27$$
$$\max\{2z_1, 3z_1\} + \max\{-3y_2, -2y_2\} = -4$$
$$\max\{w_1, 4w_1\} + \max\{-4x_2, -x_2\} = 28$$
$$y_1 - x_1 \geq 0, z_1 - y_1 \geq 0, w_1 - z_1 \geq 0$$
$$y_2 - x_2 \geq 0, z_2 - y_2 \geq 0, w_2 - z_2 \geq 0$$

**Step 4** Using Step 4 of the method, presented in Sect. 3.3, the fuzzy linear programming problem, obtained in Step 3, can be written as:

Maximize $\Re(\min\{-2w_1, 6x_1\} + \min\{2x_2, 8x_2\}, \min\{-z_1, 3y_1\} + \min\{4y_2, 6y_2\},$
$\max\{-y_1, 3z_1\} + \max\{4z_2, 6z_2\}, \max\{-2x_1, 6w_1\} + \max\{2w_2, 8w_2\})$
subject to

$$\min\{-3w_1, 7x_1\} + \min\{-2w_2, 7x_2\} = -42$$
$$\min\{2y_1, 5y_1\} + \min\{-z_2, 5y_2\} = -17$$
$$\max\{2z_1, 5z_1\} + \max\{-y_2, 5z_2\} = 45$$
$$\max\{7w_1, -3x_1\} + \max\{-2x_2, 7w_2\} = 91$$
$$\min\{x_1, 4x_1\} + \min\{-4w_2, -w_2\} = -48$$
$$\min\{2y_1, 3y_1\} + \min\{-3z_2, -2z_2\} = -27$$
$$\max\{2z_1, 3z_1\} + \max\{-3y_2, -2y_2\} = -4$$
$$\max\{w_1, 4w_1\} + \max\{-4x_2, -x_2\} = 28$$
$$y_1 - x_1 \geq 0, z_1 - y_1 \geq 0, w_1 - z_1 \geq 0$$
$$y_2 - x_2 \geq 0, z_2 - y_2 \geq 0, w_2 - z_2 \geq 0$$

**Step 5** Using Steps 7 and 8 of the method, presented in Sect. 3.3, the problem, obtained in Step 4, can be converted into the following crisp non-linear programming problem:

Maximize $\frac{1}{4}(-w_1 + 3x_1 - |-w_1 - 3x_1| + 5x_2 - |-3x_2| - \frac{1}{2}z_1 + \frac{3}{2}y_1 - |-\frac{1}{2}z_1-$

$\frac{3}{2}y_1| + 5y_2 - |-y_2| - \frac{1}{2}y_1 + \frac{3}{2}z_1 + |-\frac{1}{2}y_1 - \frac{3}{2}z_1| + 5z_2 + |-z_2| - x_1 + 3w_1 + |-$

$x_1 - 3w_1| + 5w_2 + |-3w_2|)$

subject to

$$-\frac{3}{2}w_1 + \frac{7}{2}x_1 - |-\frac{3}{2}w_1 - \frac{7}{2}x_1| - w_2 + \frac{7}{2}x_2 - |-w_2 - \frac{7}{2}x_2| = -42$$
$$\frac{7}{2}y_1 - \frac{3}{2}|y_1| - \frac{1}{2}z_2 + \frac{5}{2}y_2 - |-\frac{1}{2}z_2 - \frac{5}{2}y_2| = -17$$
$$\frac{7}{2}z_1 + \frac{3}{2}|z_1| - \frac{1}{2}y_2 + \frac{5}{2}z_2 + |-\frac{1}{2}y_2 - \frac{5}{2}z_2| = 45$$

$$\frac{7}{2}w_1 - \frac{3}{2}x_1 + |\frac{7}{2}w_1 + \frac{3}{2}x_1| - x_2 + \frac{7}{2}w_2 + |-x_2 - \frac{7}{2}w_2| = 91$$

$$\frac{5}{2}x_1 - \frac{3}{2}|x_1| - \frac{5}{2}w_2 - \frac{3}{2}|w_2| = -48$$

$$\frac{5}{2}y_1 - \frac{1}{2}|y_1| - \frac{5}{2}z_2 - \frac{1}{2}|z_2| = -27$$

$$\frac{5}{2}z_1 + \frac{1}{2}|z_1| - \frac{5}{2}y_2 + \frac{1}{2}|y_2| = -4$$

$$\frac{5}{2}w_1 + \frac{3}{2}|w_1| - \frac{5}{2}x_2 + \frac{3}{2}|x_2| = 28$$

$$y_1 - x_1 \geq 0, z_1 - y_1 \geq 0, w_1 - z_1 \geq 0$$

$$y_2 - x_2 \geq 0, z_2 - y_2 \geq 0, w_2 - z_2 \geq 0$$

**Step 6** The optimal solution of the crisp non-linear programming problem, obtained in Step 5, is $x_1 = -2, y_1 = -2, z_1 = 2, w_1 = 3, x_2 = -4, y_2 = 5, z_2 = 7$ and $w_2 = 10$.

**Step 7** Putting the values of $x_1, y_1, z_1, w_1, x_2, y_2, z_2$ and $w_2$ in $\tilde{x}_1 = (x_1, y_1, z_1, w_1)$ and $\tilde{x}_2 = (x_2, y_2, z_2, w_2)$, the exact fuzzy optimal solution is $\tilde{x}_1 = (-2, -2, 2, 3), \tilde{x}_2 = (-4, 5, 7, 10)$.

**Step 8** Putting the values of $\tilde{x}_1$ and $\tilde{x}_2$, obtained from Step 7, in the objective function the fuzzy optimal value is $(-44, 14, 48, 98)$.

### 3.4.1.3 Fuzzy Optimal Solution of the Fully Fuzzy Linear Programming Problem Chosen in Example 2.1

The fuzzy optimal solution of the fully fuzzy linear programming problem, chosen in Example 2.1, can also be obtained by using the following steps:

**Step 1** Assuming $\tilde{x}_1 = (x_1, y_1, z_1, w_1)$ and $\tilde{x}_2 = (x_2, y_2, z_2, w_2)$ the fully fuzzy linear programming problem, chosen in Example 2.1, can be written as:

Maximize $((1, 2, 3, 4) \otimes (x_1, y_1, z_1, w_1) \oplus (2, 3, 4, 5) \otimes (x_2, y_2, z_2, w_2))$

subject to

$(0, 1, 2, 3) \otimes (x_1, y_1, z_1, w_1) \oplus (1, 2, 3, 4) \otimes (x_2, y_2, z_2, w_2) = (2, 10, 24, 44)$

$(1, 2, 3, 4) \otimes (x_1, y_1, z_1, w_1) \oplus (0, 1, 2, 3) \otimes (x_2, y_2, z_2, w_2) = (1, 8, 21, 40)$

where $(x_1, y_1, z_1, w_1)$ and $(x_2, y_2, z_2, w_2)$ are non-negative trapezoidal fuzzy numbers.

**Step 2** Using the product, presented in Sect. 3.2, the fully fuzzy linear programming problem, obtained in Step 1, can be written as:

Maximize $((\min\{x_1, 4x_1\}, \min\{2y_1, 3y_1\}, \max\{2z_1, 3z_1\}, \max\{w_1, 4w_1\}) \oplus (\min\{2x_2,$

$5x_2\}, \min\{3y_2, 4y_2\}, \max\{3z_2, 4z_2\}, \max\{2w_2, 5w_2\}))$

subject to

$(\min\{0x_1, 3x_1\}, \min\{y_1, 2y_1\}, \max\{z_1, 2z_1\}, \max\{0w_1, 3w_1\}) \oplus (\min\{x_2, 4x_2\}, \min\{2y_2,$
$3y_2\}, \max\{2z_2, 3z_2\}, \min\{w_2, 4w_2\}) = (2, 10, 24, 44)$

$(\min\{x_1, 4x_1\}, \min\{2y_1, 3y_1\}, \max\{2z_1, 3z_1\}, \max\{w_1, 4w_1\}) \oplus (\min\{0x_2, 3x_2\}, \min\{y_2,$
$2y_2\}, \max\{z_2, 2z_2\}, \max\{0w_2, 3w_2\}) = (1, 8, 21, 40)$

where $(x_1, y_1, z_1, w_1)$ and $(x_2, y_2, z_2, w_2)$ are non-negative trapezoidal fuzzy numbers.

**Step 3** Using the arithmetic operations, defined in Sect. 2.1.2.1 and Definition 2.12, the fully fuzzy linear programming problem, obtained in Step 2, can be written as:

Maximize $(\min\{x_1, 4x_1\} + \min\{2x_2, 5x_2\}, \min\{2y_1, 3y_1\} + \min\{3y_2, 4y_2\},$
$\min\{2z_1, 3z_1\} + \min\{3z_2, 4z_2\}, \min\{w_1, 4w_1\} + \min\{2w_2, 5w_2\})$

subject to

$$\min\{0x_1, 3x_1\} + \min\{x_2, 4x_2\} = 2$$
$$\min\{y_1, 2y_1\} + \min\{2y_2, 3y_2\} = 10$$
$$\max\{z_1, 2z_1\} + \max\{2z_2, 3z_2\} = 24$$
$$\max\{0w_1, 3w_1\} + \max\{w_2, 4w_2\} = 44$$
$$\min\{x_1, 4x_1\} + \min\{0x_2, 3x_2\} = 1$$
$$\min\{2y_1, 3y_1\} + \min\{y_2, 2y_2\} = 8$$
$$\max\{2z_1, 3z_1\} + \max\{z_2, 2z_2\} = 21$$
$$\max\{w_1, 4w_1\} + \max\{0w_2, 3w_2\} = 40$$
$$x_1 \geq 0,\ y_1 - x_1 \geq 0,\ z_1 - y_1 \geq 0,\ w_1 - z_1 \geq 0$$
$$x_2 \geq 0,\ y_2 - x_2 \geq 0,\ z_2 - y_2 \geq 0,\ w_2 - z_2 \geq 0$$

**Step 4** Using Step 4 of the method, presented in Sect. 3.3, the fuzzy linear programming problem, obtained in Step 3, can be written as:

Maximize $\Re(\min\{x_1, 4x_1\} + \min\{2x_2, 5x_2\}, \min\{2y_1, 3y_1\} + \min\{3y_2, 4y_2\},$
$\max\{2z_1, 3z_1\} + \max\{3z_2, 4z_2\}, \max\{w_1, 4w_1\} + \max\{2w_2, 5w_2\})$

subject to

$$\min\{0x_1, 3x_1\} + \min\{x_2, 4x_2\} = 2$$
$$\min\{y_1, 2y_1\} + \min\{2y_2, 3y_2\} = 10$$
$$\max\{z_1, 2z_1\} + \max\{2z_2, 3z_2\} = 24$$
$$\max\{0w_1, 3w_1\} + \max\{w_2, 4w_2\} = 44$$
$$\min\{x_1, 4x_1\} + \max\{0x_2, 3x_2\} = 1$$
$$\min\{2y_1, 3y_1\} + \max\{y_2, 2y_2\} = 8$$
$$\max\{2z_1, 3z_1\} + \max\{z_2, 2z_2\} = 21$$

$$\max\{w_1, 4w_1\} + \max\{0w_2, 3w_2\} = 40$$
$$x_1 \geq 0, y_1 - x_1 \geq 0, z_1 - y_1 \geq 0, w_1 - z_1 \geq 0$$
$$x_2 \geq 0, y_2 - x_2 \geq 0, z_2 - y_2 \geq 0, w_2 - z_2 \geq 0$$

**Step 5** Using Steps 7 and 8 of the method, presented the problem, obtained in Step 3, can be written as:

Maximize $\dfrac{1}{4}(\dfrac{5x_1}{2} - |-\dfrac{3x_1}{2}| + \dfrac{7x_2}{2} - |-\dfrac{3x_2}{2}| + \dfrac{5y_1}{2} - |-\dfrac{y_1}{2}| + \dfrac{7y_2}{2} - |-$

$\dfrac{y_2}{2}| + \dfrac{5z_1}{2} + |-\dfrac{z_1}{2}| + \dfrac{7z_2}{2} + |-\dfrac{z_2}{2}| + \dfrac{5w_1}{2} + |-\dfrac{3w_1}{2}| + \dfrac{7z_2}{2} + |-\dfrac{3z_2}{2}|)$

subject to

$$\dfrac{3x_1}{2} - |-\dfrac{3x_1}{2}| + \dfrac{5x_2}{2} - |-\dfrac{3x_2}{2}| = 2$$

$$\dfrac{3y_1}{2} - |-\dfrac{y_1}{2}| + \dfrac{5y_2}{2} - |-\dfrac{y_2}{2}| = 10$$

$$\dfrac{3z_1}{2} + |-\dfrac{z_1}{2}| + \dfrac{5z_2}{2} + |-\dfrac{z_2}{2}| = 24$$

$$\dfrac{3w_1}{2} - |-\dfrac{3w_1}{2}| + \dfrac{5w_2}{2} + |-\dfrac{3w_2}{2}| = 44$$

$$\dfrac{5x_1}{2} - |-\dfrac{3x_1}{2}| + \dfrac{3x_2}{2} - |-\dfrac{3x_2}{2}| = 1$$

$$\dfrac{5y_1}{2} - |-\dfrac{y_1}{2}| + \dfrac{3y_2}{2} - |-\dfrac{y_2}{2}| = 8$$

$$\dfrac{3z_1}{2} + |-\dfrac{z_1}{2}| + \dfrac{3z_2}{2} + |-\dfrac{z_2}{2}| = 21$$

$$\dfrac{5w_1}{2} - |-\dfrac{3w_1}{2}| + \dfrac{3w_2}{2} - |-\dfrac{3w_2}{2}| = 40$$

$$x_1 \geq 0, y_1 - x_1 \geq 0, z_1 - y_1 \geq 0, w_1 - z_1 \geq 0$$
$$x_2 \geq 0, y_2 - x_2 \geq 0, z_2 - y_2 \geq 0, w_2 - z_2 \geq 0$$

**Step 6** Since, $x_1 \geq 0, y_1 \geq 0, z_1 \geq 0, w_1 \geq 0, x_2 \geq 0, y_2 \geq 0, z_2 \geq 0$ and $w_2 \geq 0$ so the problem, obtained in Step 5, can be written as:

Maximize $\dfrac{1}{4}(x_1 + 2x_2 + 2y_1 + 3y_2 + 3z_1 + 4z_2 + 4w_1 + 5w_2)$

subject to

$$0x_1 + x_2 = 2$$
$$x_1 + 0x_2 = 1$$
$$y_1 + 2y_2 = 10$$
$$2y_1 + y_2 = 8$$

$$2z_1 + 3z_2 = 24$$
$$3z_1 + 2z_2 = 21$$
$$3w_1 + 4w_2 = 44$$
$$4w_1 + 3w_2 = 40$$
$$x_1 \geq 0, \, y_1 - x_1 \geq 0, \, z_1 - y_1 \geq 0, \, w_1 - z_1 \geq 0$$
$$x_2 \geq 0, \, y_2 - x_2 \geq 0, \, z_2 - y_2 \geq 0, \, w_2 - z_2 \geq 0$$

**Step 7** The optimal solution of the crisp linear programming problem, obtained in Step 6, is $x_1 = 1, \, y_1 = 2, \, z_1 = 3, \, w_1 = 4, \, x_2 = 2, \, y_2 = 4, \, z_2 = 6$ and $w_2 = 8$.

**Step 8** Putting the values of $x_1, \, y_1, \, z_1, \, w_1, \, x_2, \, y_2, \, z_2$ and $w_2$ in $\tilde{x}_1 = (x_1, y_1, z_1, w_1)$ and $\tilde{x}_2 = (x_2, y_2, z_2, w_2)$, the exact fuzzy optimal solution is $\tilde{x}_1 = (1, 2, 3, 4)$, $\tilde{x}_2 = (2, 4, 6, 8)$.

**Step 9** Putting the values of $\tilde{x}_1$ and $\tilde{x}_2$, obtained from Step 8, in the objective function the fuzzy optimal value is $(5, 16, 33, 56)$.

## 3.5  Advantages of Kaur and Kumar's Method

In this section, the advantages of the Kaur and Kumar's method, presented in Sect. 3.3, over the method, presented in Chap. 2, are discussed.

(i)  The fully fuzzy linear programming problems (2.1) and (2.9) which can be solved by using the the method, presented in Chap. 2, can also be solved by using the method presented in this chapter and the obtained results are same.

(ii)  The fully fuzzy linear programming problems (3.1) and (3.2) which cannot be solved by using the method, presented in Chap. 2, can be solved by using the method presented in this chapter.

## 3.6  Real Life Application of Kaur and Kumar's Method

Kaur and Kumar [3] proposed a new method, based on tabular representation of transportation problems, to find the fuzzy optimal solution of uncapacitated fully fuzzy transportation problems and solved the real life uncapacitated fully fuzzy transportation problem, chosen in Sect. 3.6.1, to show the application of the proposed method.

In the existing method [3], it is assumed that the rank $(\frac{x_{ij} + 2y_{ij} + z_{ij}}{4})$ of fuzzy variables $\tilde{x}_{ij} = (x_{ij}, y_{ij}, z_{ij})$, representing the optimal amount of the product that should be transported from $i$th source to $j$th destination, should be greater than or equal to zero. For several unrestricted triangular fuzzy numbers, the rank can be greater than or equal to zero e.g., the rank of unrestricted triangular fuzzy number $(-2, -1, 6)$

is positive. So, to find the fuzzy optimal solution of the chosen real life problem by using its fuzzy linear programming formulation, there is a need to solve a fully fuzzy linear programming problem with unrestricted fuzzy variables.

As discussed in Sect. 3.1, there is no method in the literature to solve fully fuzzy linear programming problems with unrestricted fuzzy variables. So, the real life problem, chosen in Sect. 3.6.1, cannot be solved by any of the existing methods. Although, the uncapacitated fully fuzzy transportation problem can be solved by using the tabular method [3]. However, there may exist several problems which can be solved only by using fuzzy linear programming approach e.g., there is no tabular method in the literature to solve the crisp capacitated minimal cost flow problems which can be extended to propose a new method for solving fully fuzzy capacitated minimal cost flow problems but such problems can be formulated as a fuzzy linear programming problem and then can be solved by using the proposed method.

In this section, to show the application of the method, presented in Sect. 3.3, the real life problem, chosen in Sect. 3.6.1, is solved and it is concluded that the results, obtained by using the existing method [3] and the method, presented in Sect. 3.3, are same.

## 3.6.1 Description of the Problem

Dali Company is the leading producer of soft drinks and low-temperature foods in Taiwan. Currently, Dali plans to develop the South-East Asian market and broaden the visibility of Dali products in the Chinese market. Notably, following the entry of Taiwan to the World Trade Organization, Dali plans to seek strategic alliance with prominent international companies and introduced international bread to lighten the embedded future impact. In the domestic soft drinks market, Dali produces tea beverages to meet demand from four distribution centers in Taichung, Chiayi, Kaohsiung and Taipei, with production being based at three plants in Changhua, Touliu and Hsinchu. According to the preliminary environmental information, Table 3.1 summarizes the potential supply available from these three plants, the forecast demand from the four distribution centers and the unit transportation costs for each route used by Dali for the upcoming season.

**Table 3.1** Summarized data in the Dali case (in U.S. dollar)

| Source | Destination | | | | Supply (000 dozen bottles) |
|---|---|---|---|---|---|
| | Taichung | Chiayi | Kaohsiung | Taipei | |
| Changhua | ($8, $10, $10.8) | ($20.4, $22, $24) | ($8, $10, $10.6) | ($18.8, $20, $22) | (7.2, 8, 8.8) |
| Touliu | ($14, $15, $16) | ($18.2, $20, $22) | ($10, $12, $13) | ($6, $8, $8.8) | (12, 14, 16) |
| Hsinchu | ($18.4, $20, $21) | ($9.6, $12, $13) | ($7.8, $10, $10.8) | ($14, $15, $16) | (10.2, 12, 13.8) |
| Demand (000 dozen bottles) | (6.2, 7, 7.8) | (8.9, 10, 11.1) | (6.5, 8, 9.5) | (7.8, 9, 10.2) | |

The environmental coefficients and related parameters generally are imprecise numbers with triangular possibility distributions over the planning horizon due to incomplete or unobtainable information. For example, the available supply of the Changhua plant is ($7.2, $8, $8.8) thousand dozen bottles, the forecast demand of the Taichung distribution center is ($6.2, $7, $7.8) thousand dozen bottles and the transportation cost per dozen bottles from Changhua to Taichung is ($8, $10, $10.8).

Due to transportation costs being a major expense, the management of Dali is initiating a study to reduce these costs as much as possible.

**Solution**: The chosen real life problem [3] can be formulated into the following fully fuzzy linear programming problem:

Minimize (($8, $10, $10.8) $\otimes \tilde{x}_{11} \oplus$ ($20.4, $22, $24) $\otimes \tilde{x}_{12} \oplus$ ($8, $10, $10.6)$\otimes$
$\tilde{x}_{13} \oplus$ ($18.8, $20, $22) $\otimes \tilde{x}_{14} \oplus$ ($14, $15, $16) $\otimes \tilde{x}_{21} \oplus$ ($18.2, $20, $22)$\otimes$
$\tilde{x}_{22} \oplus$ ($10, $12, $13) $\otimes \tilde{x}_{23} \oplus$ ($6, $8, $8.8) $\otimes \tilde{x}_{24} \oplus$ ($18.4, $20, $21) $\otimes \tilde{x}_{31} \oplus$
($9.6, $12, $13) $\otimes \tilde{x}_{32} \oplus$ ($7.8, $10, $10.8) $\otimes \tilde{x}_{33} \oplus$ ($14, $15, $16) $\otimes \tilde{x}_{34}$)
subject to

$$\tilde{x}_{11} \oplus \tilde{x}_{12} \oplus \tilde{x}_{13} \oplus \tilde{x}_{14} = (7.2, 8, 8.8)$$
$$\tilde{x}_{21} \oplus \tilde{x}_{22} \oplus \tilde{x}_{23} \oplus \tilde{x}_{24} = (12, 14, 16)$$
$$\tilde{x}_{31} \oplus \tilde{x}_{32} \oplus \tilde{x}_{33} \oplus \tilde{x}_{34} = (10.2, 12, 13.8)$$
$$\tilde{x}_{11} \oplus \tilde{x}_{21} \oplus \tilde{x}_{31} = (6.2, 7, 7.8)$$
$$\tilde{x}_{12} \oplus \tilde{x}_{22} \oplus \tilde{x}_{32} = (8.9, 10, 11.1)$$
$$\tilde{x}_{13} \oplus \tilde{x}_{23} \oplus \tilde{x}_{33} = (6.5, 8, 9.5)$$
$$\tilde{x}_{14} \oplus \tilde{x}_{24} \oplus \tilde{x}_{34} = (7.8, 9, 10.2)$$
$$\tilde{x}_{ij} \succeq \tilde{0} \ \forall i = 1, \ldots, 3, \ j = 1, \ldots, 4$$

$\tilde{x}_{ij} \succeq \tilde{0} \Rightarrow \tilde{x}_{ij}$ is an unrestricted fuzzy number.

Assuming $\tilde{x}_{ij} = (x_{ij}, y_{ij}, z_{ij})$ and applying the presented method, the obtained minimum total fuzzy transportation cost is ($\$\frac{20439}{100}$, $\$\frac{7047}{20}$, $\$\frac{45651}{100}$).

*Remark 3.1* In all the existing methods [4–10] and also in the method presented in this chapter, it is assumed that if a fuzzy number $\tilde{A}$ is the fuzzy optimal solution of a problem and if there exist any fuzzy number $\tilde{B}$ such that $\Re(\tilde{A}) = \Re(\tilde{B})$ then $\tilde{B}$ will also be a fuzzy optimal solution of the same problem. Although, the fuzzy numbers, representing the minimum total fuzzy transportation cost, for the real life problem, chosen in Sect. 3.6.1, obtained by using the existing method [3] and the method presented in this chapter are ($\$\frac{2796}{10}$, $352, $382) and ($\$\frac{20439}{100}$, $\$\frac{7047}{20}$, $\$\frac{45651}{100}$) respectively but $\Re(\$\frac{2796}{10}$, $352, $382) = $\Re(\$\frac{20439}{100}$, $\$\frac{7047}{20}$, $\$\frac{45651}{100}$) which implies that there exist alternative fuzzy optimal solution of the chosen real life problem.

**Table 3.2** Results of the chosen fully fuzzy linear programming problems

| Example | Fuzzy optimal value | |
|---|---|---|
| | Method presented in Chap. 2 | Method presented in this chapter |
| 2.1 | (5, 16, 33, 56) | (5, 16, 33, 56) |
| 2.2 | (9, 27, 75, 111) | (9, 27, 75, 111) |
| 3.1 | Not applicable | $(-10, 2, \frac{831}{25}, 68)$ |
| 3.2 | Not applicable | $(-44, 14, 48, 98)$ |
| Real life problem | Not applicable | $(\$\frac{20439}{100}, \$\frac{7047}{20}, \$\frac{45651}{100})$ |

## 3.7 Comparative Study

The results of the chosen fully fuzzy linear programming problems, obtained by using the method, presented in Chap. 2, and the method presented in this chapter, are shown in Table 3.2.

The results, presented in Table 3.2, can be explained as follows:

(i) In the problems, chosen in Examples 2.1 and 2.2, all the decision variables are represented by non-negative trapezoidal fuzzy numbers. While, in the problems, chosen in Examples 3.1, 3.2 and existing real life problem [3] all the decision variables are represented by unrestricted fuzzy numbers. So, as discussed in Sect. 3.1, the problems, chosen in Examples 2.1 and 2.2, can be solved by using the method presented in Chap. 2 but the problems, chosen in Examples 3.1, 3.2 and existing real life problem [3], cannot be solved by using the method presented in Chap. 2.

(ii) Since, the method presented in this chapter can be used to find the fuzzy optimal solution of fully fuzzy linear programming problems with unrestricted fuzzy parameters. So, all the chosen problems as well as the existing real life problem [3] can be solved by using the method presented in this chapter.

## 3.8 Conclusions

On the basis of present study, it can be concluded that it is better to use the method, presented in this chapter, as compared to the method, presented in previous chapter, for solving fully fuzzy linear programming problems with equality constraints.

# References

1. Kaur, J., Kumar, A.: Exact fuzzy optimal solution of fully fuzzy linear programming problems with unrestricted fuzzy variables. Appl. Intell. **37**, 145–154 (2012)
2. Taha, H.A.: Operations Research: An Introduction. Prentice-Hall, New Jersey (2003)
3. Kaur, A., Kumar, A.: A new method for solving fuzzy transportation problems using ranking function. Appl. Math. Model. **35**, 5652–5661 (2011)
4. Maleki, H.R., Tata, M., Mashinchi, M.: Linear programming with fuzzy variables. Fuzzy Sets Syst. **109**, 21–33 (2000)
5. Maleki, H.R.: Ranking functions and their applications to fuzzy linear programming. Far East J. Math. Sci. **4**, 283–301 (2002)
6. Ebrahimnejad, A., Nasseri, S.H., Lotfi, F.H., Soltanifar, M.: A primal-dual method for linear programming problems with fuzzy variables. Eur. J. Ind. Eng. **4**, 189–209 (2010)
7. Buckley, J., Feuring, T.: Evolutionary algorithm solution to fuzzy problems: fuzzy linear programming. Fuzzy Sets Syst. **109**, 35–53 (2000)
8. Hashemi, S.M., Modarres, M., Nasrabadi, E., Nasrabadi, M.M.: Fully fuzzified linear programming, solution and duality. J. Intell. Fuzzy Syst. **17**, 253–261 (2006)
9. Allahviranloo, T., Lotfi, F.H., Kiasary, M.K., Kiani, N.A., Alizadeh, L.: Solving fully fuzzy linear programming problem by the ranking functhion. Appl. Math. Sci. **2**, 19–32 (2008)
10. Lotfi, F.H., Allahviranloo, T., Jondabeha, M.A., Alizadeh, L.: Solving a fully fuzzy linear programming using lexicography method and fuzzy approximate solution. Appl. Math. Model. **33**, 3151–3156 (2009)

# Chapter 4
# Fuzzy Optimal Solution of Fully Fuzzy Linear Programming Problems with Equality Constraints Having LR Flat Fuzzy Numbers

In the literature [1] it is pointed out that there may exist several real life problems in which it is not always possible to represent all the parameters as triangular or trapezoidal fuzzy numbers and due to the same reason several authors [1–5] have represented the parameters as *LR* flat fuzzy numbers instead of triangular or trapezoidal fuzzy numbers.

To the best of our knowledge, till now no one have defined the product of such *LR* fuzzy numbers or *LR* flat fuzzy numbers which are neither non-negative nor non-positive. Due to non-existence of such product, there was no method in the literature for solving such fully fuzzy linear programming problems in which some or all the parameters are represented by such *LR* fuzzy numbers or *LR* flat fuzzy numbers which are neither non-negative nor non-positive. Therefore, Kaur and Kumar [6] proposed the product of such *LR* flat fuzzy numbers which are neither non-negative nor non-positive and in this chapter, the product of such fuzzy numbers is presented and also the limitations of the method, presented in Chap. 3, are pointed out. To overcome the limitations of the method, presented in Chap. 3, a method proposed by Kaur and Kumar [6], is presented to find the fuzzy optimal solution of fully fuzzy linear programming problems with equality constraints.

## 4.1 Preliminaries

In this section, some basic definitions and arithmetic operations of *LR* flat fuzzy numbers are presented [1].

The contents of this chapter are published in *Applied Mathematical Modelling* 37 (2013) 7142–7153.

55
J. Kaur and A. Kumar, *An Introduction to Fuzzy Linear Programming Problems*, Studies in Fuzziness and Soft Computing 340, DOI 10.1007/978-3-319-31274-3_4

### 4.1.1  Basic Definitions

In this section, some basic definitions are presented.

**Definition 4.1** [1] A function $L : [0, \infty) \rightarrow [0, 1]$ (or $R : [0, \infty) \rightarrow [0, 1]$) is said to be reference function of fuzzy number if and only if
(i) $L(0) = 1$ (or $R(0) = 1$)
(ii) $L$ (or $R$) is non-increasing on $[0, \infty)$.

**Definition 4.2** [1] A fuzzy number $\tilde{A}$, defined on universal set of real numbers $\mathbb{R}$, denoted as $(m, n, \alpha, \beta)_{LR}$, is said to be an $LR$ flat fuzzy number if its membership function $\mu_{\tilde{A}}(x)$ is given by

$$\mu_{\tilde{A}}(x) = \begin{cases} L(\frac{m-x}{\alpha}) & x \leq m, \ \alpha > 0 \\ R(\frac{x-n}{\beta}) & x \geq n, \ \beta > 0 \\ 1 & m \leq x \leq n \end{cases}$$

**Definition 4.3** [1] Let $\tilde{A} = (m, n, \alpha, \beta)_{LR}$ be an $LR$ flat fuzzy number and $\lambda$ be a real number in the interval $[0, 1]$. Then, the crisp set $A^\lambda = \{x \in X : \mu_{\tilde{A}}(x) \geq \lambda\} = [m - \alpha L^{-1}(\lambda), n + \beta R^{-1}(\lambda)]$, is said to be $\lambda$-cut of $\tilde{A}$.

**Definition 4.4** [1] An $LR$ flat fuzzy number $\tilde{A} = (m, n, \alpha, \beta)_{LR}$ is said to be zero $LR$ flat fuzzy number if and only if $m = 0, n = 0, \alpha = 0$ and $\beta = 0$.

**Definition 4.5** [1] Two $LR$ flat fuzzy numbers $\tilde{A}_1 = (m_1, n_1, \alpha_1, \beta_1)_{LR}$ and $\tilde{A}_2 = (m_2, n_2, \alpha_2, \beta_2)_{LR}$ are said to be equal i.e., $\tilde{A}_1 = \tilde{A}_2$ if and only if $m_1 = m_2, n_1 = n_2, \alpha_1 = \alpha_2$ and $\beta_1 = \beta_2$.

**Definition 4.6** [7] An $LR$ flat fuzzy number $\tilde{A} = (m, n, \alpha, \beta)_{LR}$ is said to be non-negative $LR$ flat fuzzy number if and only if $m - \alpha \geq 0$ and is said to be non-positive $LR$ flat fuzzy number if and only if $m - \alpha \leq 0$.

**Definition 4.7** An $LR$ flat fuzzy number $\tilde{A} = (m, n, \alpha, \beta)_{LR}$ is said to be unrestricted $LR$ flat fuzzy number if and only if $m - \alpha$ is a real number.

*Remark 4.1* If $m = n$ then an $LR$ flat fuzzy number $(m, n, \alpha, \beta)_{LR}$ is said to be an $LR$ fuzzy number and is denoted as $(m, m, \alpha, \beta)_{LR}$ or $(n, n, \alpha, \beta)_{LR}$ or $(m, \alpha, \beta)_{LR}$ or $(n, \alpha, \beta)_{LR}$.

*Remark 4.2* If $m = n$ and $L(x) = R(x) = \max\{0, 1 - x\}$ then an $LR$ flat fuzzy number $(m, n, \alpha, \beta)_{LR}$ is said to be a triangular fuzzy number and is denoted as $(a, b, c)$ where $a = m - \alpha, b = m(orn), c = m + \beta(or\ n + \beta)$.

*Remark 4.3* If $m \neq n$ and $L(x) = R(x) = \max\{0, 1 - x\}$ then an $LR$ flat fuzzy number $(m, n, \alpha, \beta)_{LR}$ is said to be a trapezoidal fuzzy number and is denoted as $(a, b, c, d)$ where $a = m - \alpha, b = m, c = n, d = n + \beta$.

### 4.1.2 Arithmetic Operations

In this section, the arithmetic operations between $LR$ flat fuzzy numbers are presented [1].

Let $\tilde{A}_1 = (m_1, n_1, \alpha_1, \beta_1)_{LR}$, $\tilde{A}_2 = (m_2, n_2, \alpha_2, \beta_2)_{LR}$ be any $LR$ flat fuzzy numbers and $\tilde{A}_3 = (m_3, n_3, \alpha_3, \beta_3)_{RL}$ be any $RL$ flat fuzzy number. Then,

(i)  $\tilde{A}_1 \oplus \tilde{A}_2 = (m_1 + m_2, n_1 + n_2, \alpha_1 + \alpha_2, \beta_1 + \beta_2)_{LR}$

(ii)  $\tilde{A}_1 \ominus \tilde{A}_3 = (m_1 - n_3, n_1 - m_3, \alpha_1 + \beta_3, \beta_1 + \alpha_3)_{LR}$

(iii)  If $\tilde{A}_1$ and $\tilde{A}_2$ both are non-negative, then
$\tilde{A}_1 \otimes \tilde{A}_2 = (m_1 m_2, n_1 n_2, m_1 \alpha_2 + \alpha_1 m_2 - \alpha_1 \alpha_2, n_1 \beta_2 + \beta_1 n_2 + \beta_1 \beta_2)_{LR}$

(iv)  If $\tilde{A}_1$ is non-positive and $\tilde{A}_2$ is non-negative, then
$\tilde{A}_1 \otimes \tilde{A}_2 = (m_1 n_2, n_1 m_2, \alpha_1 n_2 - m_1 \beta_2 + \alpha_1 \beta_2, \beta_1 m_2 - n_1 \alpha_2 - \beta_1 \alpha_2)_{LR}$

(v)  If $\tilde{A}_1$ is non-negative and $\tilde{A}_2$ is non-positive, then
$\tilde{A}_1 \otimes \tilde{A}_2 = (n_1 m_2, m_1 n_2, n_1 \alpha_2 - \beta_1 m_2 + \beta_1 \alpha_2, m_1 \beta_2 - \alpha_1 n_2 - \alpha_1 \beta_2)_{LR}$

(vi)  If $\tilde{A}_1$ and $\tilde{A}_2$ both are non-positive, then
$\tilde{A}_1 \otimes \tilde{A}_2 = (n_1 n_2, m_1 m_2, -n_1 \beta_2 - \beta_1 n_2 - \beta_1 \beta_2, -m_1 \alpha_2 - \alpha_1 m_2 + \alpha_1 \alpha_2)_{LR}$

(vii)  $\lambda \tilde{A}_1 = \begin{cases} (\lambda m_1, \lambda n_1, \lambda \alpha_1, \lambda \beta_1)_{LR} & \lambda \geq 0 \\ (\lambda n_1, \lambda m_1, -\lambda \beta_1, -\lambda \alpha_1)_{RL} & \lambda \leq 0 \end{cases}$

There also exist another formula [1] for the product of such $LR$ flat fuzzy numbers in which the spreads $\alpha_1, \alpha_2, \beta_1$ and $\beta_2$ are smaller as compared to the mean values $m_1$ and $m_2$:

(i)  If $\tilde{A}_1$ and $\tilde{A}_2$ both are non-negative, then
$\tilde{A}_1 \odot \tilde{A}_2 = (m_1 m_2, n_1 n_2, m_1 \alpha_2 + \alpha_1 m_2, n_1 \beta_2 + \beta_1 n_2)_{LR}$

(ii)  If $\tilde{A}_1$ is non-positive and $\tilde{A}_2$ is non-negative, then
$\tilde{A}_1 \odot \tilde{A}_2 = (m_1 n_2, n_1 m_2, \alpha_1 n_2 - m_1 \beta_2, \beta_1 m_2 - n_1 \alpha_2)_{LR}$

(iii)  If $\tilde{A}_1$ is non-negative and $\tilde{A}_2$ is non-positive, then
$\tilde{A}_1 \odot \tilde{A}_2 = (n_1 m_2, m_1 n_2, n_1 \alpha_2 - \beta_1 m_2, m_1 \beta_2 - \alpha_1 n_2)_{LR}$

(iv)  If $\tilde{A}_1$ and $\tilde{A}_2$ both are non-positive, then
$\tilde{A}_1 \odot \tilde{A}_2 = (n_1 n_2, m_1 m_2, -n_1 \beta_2 - \beta_1 n_2, -m_1 \alpha_2 - \alpha_1 m_2)_{LR}$

*Remark 4.4* Using Yager's ranking approach [8], the value of Yager's ranking index $\Re(\tilde{A})$ for any parameter, represented by $LR$ flat fuzzy number $\tilde{A} = (m, n, \alpha, \beta)_{LR}$ is as follows:

(i) If $L(x) = R(x) = \max\{0, 1 - x\}$ then $\Re(\tilde{A}) = \frac{1}{2}(m + n) + \frac{1}{4}(\beta - \alpha)$

(ii) If $L(x) = \max\{0, 1 - x\}$ and $R(x) = \max\{0, 1 - x^2\}$ then $\Re(\tilde{A}) = \frac{1}{2}(m + n) + \frac{1}{3}\beta - \frac{1}{4}\alpha$

(iii) If $L(x) = \max\{0, 1 - x^2\}$ and $R(x) = \max\{0, 1 - x\}$ then $\Re(A) = \frac{1}{2}(m + n) + \frac{1}{4}\beta - \frac{1}{3}\alpha$

To find the ranking index for $LR$ fuzzy numbers put $m = n$ in the all of the above obtained formulas of $LR$ flat fuzzy numbers.

## 4.2   Product of Unrestricted *LR* Flat Fuzzy Numbers

The existing product rule, presented in Sect. 4.1.2 can be used to find the product of such *LR* fuzzy numbers or *LR* flat fuzzy numbers which are either non-negative or non-positive. To the best of our knowledge, till now no one have defined the product of such *LR* fuzzy numbers or *LR* flat fuzzy numbers which are neither non-negative nor non-positive e.g., if $\tilde{A}_1 = (1, 3, 4, 2)_{LR}$ and $\tilde{A}_2 = (2, 4, 5, 3)_{LR}$ are two *LR* flat fuzzy numbers then there is no product rule to find the value of $\tilde{A}_1 \otimes \tilde{A}_2$ or $\tilde{A}_1 \odot \tilde{A}_2$. Due to non-existence of such product, there was no method in the literature for solving such fully fuzzy linear programming problems in which some or all the parameters are represented by such *LR* fuzzy numbers or *LR* flat fuzzy numbers which are neither non-negative nor non-positive so Kaur and Kumar [6] proposed the product of such *LR* flat fuzzy numbers. In this section, corresponding to the existing product rules, presented in Sect. 4.1.2, new product rules, proposed by Kaur and Kumar [6], are introduced.

### 4.2.1   New Product Corresponding to the Existing Product ⊗

In this section, new product corresponding to the existing product $\otimes$, proposed by Kaur and Kumar [6], is introduced.

**Proposition 4.1** *If* $\tilde{A}_1 = (m_1, n_1, \alpha_1, \beta_1)_{LR}$ *and* $\tilde{A}_2 = (m_2, n_2, \alpha_2, \beta_2)_{LR}$ *are two LR flat fuzzy numbers such that* $m_1 - \alpha_1 < 0$ *and* $m_1 \geq 0$ *then* $\tilde{A}_1 \otimes \tilde{A}_2 = (m'_1, n'_1, \alpha'_1, \beta'_1)_{LR}$, *where* $m'_1 = \min\{m_1 m_2, n_1 m_2\}$, $n'_1 = \max\{m_1 n_2, n_1 n_2\}$, $\alpha'_1 = \min\{m_1 m_2, n_1 m_2\} - \min\{m_1 n_2 + m_1 \beta_2 - \alpha_1 n_2 - \alpha_1 \beta_2, n_1 m_2 - n_1 \alpha_2 + \beta_1 m_2 - \beta_1 \alpha_2\}$, $\beta'_1 = \max\{m_1 m_2 - m_1 \alpha_2 - \alpha_1 m_2 + \alpha_1 \alpha_2, n_1 n_2 + n_1 \beta_2 + \beta_1 n_2 + \beta_1 \beta_2\} - \max\{m_1 n_2, n_1 n_2\}$.

*Proof* Let $\tilde{A}_1 = (m_1, n_1, \alpha_1, \beta_1)_{LR}$ and $\tilde{A}_2 = (m_2, n_2, \alpha_2, \beta_2)_{LR}$ be two *LR* flat fuzzy numbers such that $m_1 - \alpha_1 < 0$ and $m_1 \geq 0$. Then, using the Definition 4.3, $A_1^\lambda = [m_1 - \alpha_1 L^{-1}(\lambda), n_1 + \beta_1 R^{-1}(\lambda)]$ and $A_2^\lambda = [m_2 - \alpha_2 L^{-1}(\lambda), n_2 + \beta_2 R^{-1}(\lambda)]$. Since $m_1 - \alpha_1 < 0$ and $m_1 \geq 0$ so $m_1 - \alpha_1 L^{-1}(\lambda) \leq 0$ for $\lambda \geq L(\frac{m_1}{\alpha_1})$ and $m_1 - \alpha_1 L^{-1}(\lambda) \geq 0$ for $\lambda \leq L(\frac{m_1}{\alpha_1})$ and $n_1 + \beta_1 R^{-1}(\lambda) \geq 0$ for all $\lambda$, so to find the product of $\tilde{A}_1$ and $\tilde{A}_2$ there is a need to consider the following five cases:

**Case (i)**   If $m_2 - \alpha_2 \geq 0$ then $m_2 - \alpha_2 L^{-1}(\lambda) \geq 0$ and $n_2 + \beta_2 R^{-1}(\lambda) \geq 0$ for all $\lambda$ so the following two subcases may arise to find the product of $A_1^\lambda$ and $A_2^\lambda$:

(a)   If $m_1 - \alpha_1 L^{-1}(\lambda) \geq 0$ then
$A_1^\lambda A_2^\lambda = [(m_1 - \alpha_1 L^{-1}(\lambda))(m_2 - \alpha_2 L^{-1}(\lambda)), (n_1 + \beta_1 R^{-1}(\lambda))(n_2 + \beta_2 R^{-1}(\lambda))]$
Putting $\lambda = 1$,
$$A_1^\lambda A_2^\lambda = [m_1 m_2, n_1 n_2]. \tag{4.1}$$

(b)  If $m_1 - \alpha_1 L^{-1}(\lambda) \leq 0$ then
$A_1^\lambda A_2^\lambda = [(m_1 - \alpha_1 L^{-1}(\lambda))(n_2 + \beta_2 R^{-1}(\lambda)), (n_1 + \beta_1 R^{-1}(\lambda))(n_2 + \beta_2 R^{-1}(\lambda))]$
Putting $\lambda = 0$,

$$A_1^\lambda A_2^\lambda = [m_1 n_2 + m_1 \beta_2 - \alpha_1 n_2 - \alpha_1 \beta_2, n_1 n_2 + n_1 \beta_2 + \beta_1 n_2 + \beta_1 \beta_2]. \tag{4.2}$$

Now combining (4.1) and (4.2):
$\tilde{A}_1 \otimes \tilde{A}_2 = (m_1'', n_1'', \alpha_1'', \beta_1'')_{LR}$
where $m_1'' = m_1 m_2, n_1'' = n_1 n_2, \alpha_1'' = m_1 m_2 - m_1 n_2 + m_1 \beta_2 - \alpha_1 n_2 - \alpha_1 \beta_2, \beta_1'' = n_1 n_2 + n_1 \beta_2 + \beta_1 n_2 + \beta_1 \beta_2 - n_1 n_2$.

**Case (ii)**  If $m_2 - \alpha_2 < 0, m_2 \geq 0$ then $m_2 - \alpha_2 L^{-1}(\lambda) \geq 0$ for $\lambda \leq L(\frac{m_2}{\alpha_2})$, $m_2 - \alpha_2 L^{-1}(\lambda) \leq 0$ for $\lambda \geq L(\frac{m_2}{\alpha_2})$ and $n_2 + \beta_2 R^{-1}(\lambda) \geq 0$ for all $\lambda$ so, the four subcases may arise to find the product of $A_1^\lambda$ and $A_2^\lambda$. Since, the aim is to find the product of $A_1^\lambda$ and $A_2^\lambda$ corresponding to $\lambda = 0$ and $\lambda = 1$ so there is a need to consider only the following two subcases:

(a)  If $m_1 - \alpha_1 L^{-1}(\lambda) \leq 0$ and $m_2 - \alpha_2 L^{-1}(\lambda) \leq 0$ then
$A_1^\lambda A_2^\lambda = [\min\{(m_1 - \alpha_1 L^{-1}(\lambda))(n_2 + \beta_2 R^{-1}(\lambda)), (n_1 + \beta_1 R^{-1}(\lambda))(m_2 - \alpha_2 L^{-1}(\lambda))\}, \max\{(m_1 - \alpha_1 L^{-1}(\lambda))(m_2 - \alpha_2 L^{-1}(\lambda)), (n_1 + \beta_1 R^{-1}(\lambda))(n_2 + \beta_2 R^{-1}(\lambda))\}]$
Putting $\lambda = 0$,

$$A_1^\lambda A_2^\lambda = [\min\{m_1 n_2 + m_1 \beta_2 - \alpha_1 n_2 - \alpha_1 \beta_2, n_1 m_2 - n_1 \alpha_2 + \beta_1 m_2 - \beta_1 \alpha_2\}, \max\{m_1 m_2 - m_1 \alpha_2 - \alpha_1 m_2 + \alpha_1 \alpha_2, n_1 n_2 + n_1 \beta_2 + \beta_1 n_2 + \beta_1 \beta_2\}]. \tag{4.3}$$

(b)  If $m_1 - \alpha_1 L^{-1}(\lambda) \geq 0$ and $m_2 - \alpha_2 L^{-1}(\lambda) \geq 0$ then
$A_1^\lambda A_2^\lambda = [(m_1 - \alpha_1 L^{-1}(\lambda))(m_2 - \alpha_2 L^{-1}(\lambda)), (n_1 + \beta_1 R^{-1}(\lambda))(n_2 + \beta_2 R^{-1}(\lambda))]$
Putting $\lambda = 1$,
$$A_1^\lambda A_2^\lambda = [m_1 m_2, n_1 n_2]. \tag{4.4}$$

Now combining (4.3) and (4.4):
$\tilde{A}_1 \otimes \tilde{A}_2 = (m_2'', n_2'', \alpha_2'', \beta_2'')_{LR}$
where $m_2'' = m_1 m_2, n_2'' = n_1 n_2, \alpha_2'' = m_1 m_2 - \min\{m_1 n_2 + m_1 \beta_2 - \alpha_1 n_2 - \alpha_1 \beta_2 - m_1 m_2, n_1 m_2 - n_1 \alpha_2 + \beta_1 m_2 - \beta_1 \alpha_2\}, \beta_2'' = \max\{m_1 m_2 - m_1 \alpha_2 - \alpha_1 m_2 + \alpha_1 \alpha_2, n_1 n_2 + n_1 \beta_2 + \beta_1 n_2 + \beta_1 \beta_2\} - n_1 n_2$.

**Case (iii)**  If $m_2 < 0, n_2 \geq 0$ then $m_2 - \alpha_2 L^{-1}(\lambda) \leq 0$ and $n_2 + \beta_2 R^{-1}(\lambda) \geq 0$ for all $\lambda$ so the following two subcases may arise to find the product of $A_1^\lambda$ and $A_2^\lambda$:

(a)  If $m_1 - \alpha_1 L^{-1}(\lambda) \geq 0$ then
$A_1^\lambda A_2^\lambda = [(n_1 + \beta_1 R^{-1}(\lambda))(m_2 - \alpha_2 L^{-1}(\lambda)), (n_1 + \beta_1 R^{-1}(\lambda))(n_2 + \beta_2 R^{-1}(\lambda))]$

Putting $\lambda = 1$,

$$A_1^\lambda A_2^\lambda = [n_1 m_2, n_1 n_2]. \tag{4.5}$$

(b) If $m_1 - \alpha_1 L^{-1}(\lambda) \leq 0$ then
$A_1^\lambda A_2^\lambda = [\min\{(m_1 - \alpha_1 L^{-1}(\lambda))(n_2 + \beta_2 R^{-1}(\lambda)), (n_1 + \beta_1 R^{-1}(\lambda))$
$(m_2 - \alpha_2 L^{-1}(\lambda))\}, \max\{(m_1 - \alpha_1 L^{-1}(\lambda))(m_2 - \alpha_2 L^{-1}(\lambda)), (n_1 +$
$\beta_1 R^{-1}(\lambda))(n_2 + \beta_2 R^{-1}(\lambda))\}]$
Putting $\lambda = 0$,

$$A_1^\lambda A_2^\lambda = [\min\{m_1 n_2 + m_1 \beta_2 - \alpha_1 n_2 - \alpha_1 \beta_2, n_1 m_2 + \beta_1 m_2$$
$$- n_1 \alpha_2 - \beta_1 \alpha_2\}, \max\{m_1 m_2 - m_1 \alpha_2 - \alpha_1 m_2 + \alpha_1 \alpha_2, n_1 n_2$$
$$+ n_1 \beta_2 + \beta_1 n_2 + \beta_1 \beta_2\}]. \tag{4.6}$$

Now combining (4.5) and (4.6):
$\tilde{A}_1 \otimes \tilde{A}_2 = (m_3'', n_3'', \alpha_3'', \beta_3'')_{LR}$
where $m_3'' = n_1 m_2, n_3'' = n_1 n_2, \alpha_3'' = n_1 m_2 - \min\{m_1 n_2 + m_1 \beta_2 - \alpha_1 n_2 - \alpha_1 \beta_2, n_1 m_2 + \beta_1 m_2 - n_1 \alpha_2 - \beta_1 \alpha_2\}, \beta_3'' = \max\{m_1 m_2 - m_1 \alpha_2 - \alpha_1 m_2 + \alpha_1 \alpha_2, n_1 n_2 + n_1 \beta_2 + \beta_1 n_2 + \beta_1 \beta_2\} - n_1 n_2.$

**Case (iv)** If $n_2 < 0, n_2 + \beta_2 \geq 0$ then $m_2 - \alpha_2 L^{-1}(\lambda) \leq 0$ for all $\lambda$ and $n_2 + \beta_2 R^{-1}(\lambda) \leq 0$ for $\lambda \leq R(-\frac{n_2}{\beta_2})$, $n_2 + \beta_2 R^{-1}(\lambda) \geq 0$ for $\lambda \geq R(-\frac{n_2}{\beta_2})$
so the four subcases may arise to find the product of $A_1^\lambda$ and $A_2^\lambda$. Since, the aim is to find the product of $A_1^\lambda$ and $A_2^\lambda$ corresponding to $\lambda = 0$ and $\lambda = 1$ so there is a need to consider only the following two subcases:

(a) If $m_1 - \alpha_1 L^{-1}(\lambda) \geq 0$ and $n_2 + \beta_2 R^{-1}(\lambda) \leq 0$ then
$A_1^\lambda A_2^\lambda = [(n_1 + \beta_1 R^{-1}(\lambda))(m_2 - \alpha_2 L^{-1}(\lambda)), (m_1 - \alpha_1 L^{-1}(\lambda))(n_2 + \beta_2 R^{-1}(\lambda))]$
Putting $\lambda = 1$,

$$A_1^\lambda A_2^\lambda = [n_1 m_2, m_1 n_2] \tag{4.7}$$

(b) If $m_1 - \alpha_1 L^{-1}(\lambda) \leq 0$ and $n_2 + \beta_2 R^{-1}(\lambda) \geq 0$ then
$A_1^\lambda A_2^\lambda = [\min\{(m_1 - \alpha_1 L^{-1}(\lambda))(n_2 + \beta_2 R^{-1}(\lambda)), (n_1 + \beta_1 R^{-1}(\lambda))$
$(m_2 - \alpha_2 L^{-1}(\lambda))\}, \max\{(m_1 - \alpha_1 L^{-1}(\lambda))(m_2 - \alpha_2 L^{-1}(\lambda)), (n_1 +$
$\beta_1 R^{-1}(\lambda))(n_2 + \beta_2 R^{-1}(\lambda))\}]$
Putting $\lambda = 0$,

$$A_1^\lambda A_2^\lambda = [\min\{m_1 n_2 + m_1 \beta_2 - \alpha_1 n_2 - \alpha_1 \beta_2, n_1 m_2 - n_1 \alpha_2$$
$$+ \beta_1 m_2 - \beta_1 \alpha_2\}, \max\{m_1 m_2 - m_1 \alpha_2 - \alpha_1 m_2 + \alpha_1 \alpha_2, n_1 n_2$$
$$+ n_1 \beta_2 + \beta_1 n_2 + \beta_1 \beta_2\}]. \tag{4.8}$$

Now combining (4.7) and (4.8):
$\tilde{A}_1 \otimes \tilde{A}_2 = (m_4'', n_4'', \alpha_4'', \beta_4'')_{LR}$
where $m_4'' = n_1 m_2, n_4'' = m_1 n_2, \alpha_4'' = n_1 m_2 - \min\{m_1 n_2 + m_1 \beta_2 -$

$\alpha_1 n_2 - \alpha_1\beta_2, n_1 m_2 - n_1\alpha_2 + \beta_1 m_2 - \beta_1\alpha_2\}, \beta_4'' = \max\{m_1 m_2 - m_1\alpha_2 - \alpha_1 m_2 + \alpha_1\alpha_2, n_1 n_2 + n_1\beta_2 + \beta_1 n_2 + \beta_1\beta_2\} - m_1 n_2.$

**Case (v)** If $n_2 + \beta_2 < 0$ then $m_2 - \alpha_2 L^{-1}(\lambda) \le 0$ and $n_2 + \beta_2 R^{-1}(\lambda) \le 0$ for all $\lambda$ so the following two subcases may arise:

(a) If $m_1 - \alpha_1 L^{-1}(\lambda) \ge 0$ then
$A_1^\lambda A_2^\lambda = [(n_1 + \beta_1 R^{-1}(\lambda))(m_2 - \alpha_2 L^{-1}(\lambda)), (m_1 - \alpha_1 L^{-1}(\lambda))(n_2 + \beta_2 R^{-1}(\lambda))]$
Putting $\lambda = 1$,

$$A_1^\lambda A_2^\lambda = [n_1 m_2, m_1 n_2]. \tag{4.9}$$

(b) If $m_1 - \alpha_1 L^{-1}(\lambda) \le 0$ then
$A_1^\lambda A_2^\lambda = [(n_1 + \beta_1 R^{-1}(\lambda))(m_2 - \alpha_2 L^{-1}(\lambda)), (m_1 - \alpha_1 L^{-1}(\lambda))(m_2 - \alpha_2 L^{-1}(\lambda))]$
Putting $\lambda = 0$,

$$A_1^\lambda A_2^\lambda = [n_1 m_2 - n_1\alpha_2 + \beta_1 m_2 - \beta_1\alpha_2, m_1 m_2 - m_1\alpha_2 - \alpha_1 m_2 + \alpha_1\alpha_2]. \tag{4.10}$$

Now combining (4.9) and (4.10):
$\tilde{A}_1 \otimes \tilde{A}_2 = (m_5'', n_5'', \alpha_5'', \beta_5'')_{LR}$
where $m_5'' = n_1 m_2, n_5'' = m_1 n_2, \alpha_5'' = n_1 m_2 - n_1 m_2 - n_1\alpha_2 + \beta_1 m_2 - \beta_1\alpha_2, \beta_5'' = m_1 m_2 - m_1\alpha_2 - \alpha_1 m_2 + \alpha_1\alpha_2 - m_1 n_2.$

Combining the results of all five cases the following result is obtained: If $\tilde{A}_1 = (m_1, n_1, \alpha_1, \beta_1)_{LR}$ and $\tilde{A}_2 = (m_2, n_2, \alpha_2, \beta_2)_{LR}$ are two *LR* flat fuzzy numbers such that $m_1 - \alpha_1 < 0, m_1 \ge 0$ and $\tilde{A}_2$ is any *LR* flat fuzzy number, then $\tilde{A}_1 \otimes \tilde{A}_2 = (m_1', n_1', \alpha_1', \beta_1')_{LR}$,
where $m_1' = \min\{m_1 m_2, n_1 m_2\}, n_1' = \max\{m_1 n_2, n_1 n_2\}, \alpha_1' = \min\{m_1 m_2, n_1 m_2\} - \min\{m_1 n_2 + m_1\beta_2 - \alpha_1 n_2 - \alpha_1\beta_2, n_1 m_2 - n_1\alpha_2 + \beta_1 m_2 - \beta_1\alpha_2\}, \beta_1' = \max\{m_1 m_2 - m_1\alpha_2 - \alpha_1 m_2 + \alpha_1\alpha_2, n_1 n_2 + n_1\beta_2 + \beta_1 n_2 + \beta_1\beta_2\} - \max\{m_1 n_2, n_1 n_2\}.$

**Proposition 4.2** *If $\tilde{A}_1 = (m_1, n_1, \alpha_1, \beta_1)_{LR}$ and $\tilde{A}_2 = (m_2, n_2, \alpha_2, \beta_2)_{LR}$ are two LR flat fuzzy numbers such that $m_1 < 0$ and $n_1 \ge 0$ then $\tilde{A}_1 \otimes \tilde{A}_2 = (m_2', n_2', \alpha_2', \beta_2')_{LR}$, where $m_2' = \min\{m_1 n_2, n_1 n_2\}, n_2' = \max\{m_1 m_2, n_1 n_2\}, \alpha_2' = \min\{m_1 n_2, n_1 n_2\} - \min\{m_1 n_2 + m_1\beta_2 - \alpha_1 n_2 - \alpha_1\beta_2, n_1 m_2 - n_1\alpha_2 + \beta_1 m_2 - \beta_1\alpha_2\}, \beta_2' = \max\{m_1 m_2 - m_1\alpha_2 - \alpha_1 m_2 + \alpha_1\alpha_2, n_1 n_2 + n_1\beta_2 + \beta_1 n_2 + \beta_1\beta_2\} - \max\{m_1 m_2, n_1 n_2\}.$*

*Proof* Similar to Proposition 4.1.

**Proposition 4.3** *If $\tilde{A}_1 = (m_1, n_1, \alpha_1, \beta_1)_{LR}$ and $\tilde{A}_2 = (m_2, n_2, \alpha_2, \beta_2)_{LR}$ are two LR flat fuzzy numbers such that $n_1 < 0$ and $n_1 + \beta_1 \ge 0$ then $\tilde{A}_1 \otimes \tilde{A}_2 = (m_3', n_3', \alpha_3', \beta_3')_{LR}$, where $m_3' = \min\{m_1 n_2, n_1 n_2\}, n_3' = \max\{n_1 m_2, m_1 m_2\}, \alpha_3' = \min\{m_1 n_2, n_1 n_2\} - \min\{m_1 n_2 + m_1\beta_2 - \alpha_1 n_2 - \alpha_1\beta_2, n_1 m_2 - n_1\alpha_2 + \beta_1 m_2 - \beta_1\alpha_2\}, \beta_3' = \max\{m_1 m_2 - m_1\alpha_2 - \alpha_1 m_2 + \alpha_1\alpha_2, n_1 n_2 + n_1\beta_2 + \beta_1 n_2 + \beta_1\beta_2\} - \max\{n_1 m_2, m_1 m_2\}.$*

*Proof* Similar to Proposition 4.1.

**Proposition 4.4** *If $\tilde{A}_1 = (m_1, n_1, \alpha_1, \beta_1)_{LR}$ and $\tilde{A}_2 = (m_2, n_2, \alpha_2, \beta_2)_{LR}$ are two LR flat fuzzy numbers such that $n_1 + \beta_1 < 0$ then $\tilde{A}_1 \otimes \tilde{A}_2 = (m_4', n_4', \alpha_4', \beta_4')_{LR}$, where $m_4' = \min\{m_1 n_2, n_1 n_2\}$, $n_4' = \max\{m_1 m_2, n_1 m_2\}$, $\alpha_4' = \min\{m_1 n_2, n_1 n_2\} - \min\{m_1 n_2 + m_1 \beta_2 - \alpha_1 n_2 - \alpha_1 \beta_2, n_1 n_2 + n_1 \beta_2 + \beta_1 n_2 + \beta_1 \beta_2\}$, $\beta_4' = \max\{n_1 m_2 - n_1 \alpha_2 + \beta_1 m_2 - \beta_1 \alpha_2, m_1 m_2 - m_1 \alpha_2 - \alpha_1 m_2 + \alpha_1 \alpha_2\} - \max\{m_1 m_2, n_1 m_2\}$.*

*Proof* Similar to Proposition 4.1.

**Proposition 4.5** *If $\tilde{A}_1 = (m_1, n_1, \alpha_1, \beta_1)_{LR}$ and $\tilde{A}_2 = (m_2, n_2, \alpha_2, \beta_2)_{LR}$ are two LR flat fuzzy numbers such that $m_1 - \alpha_1 \geq 0$ then $\tilde{A}_1 \otimes \tilde{A}_2 = (m_5', n_5', \alpha_5', \beta_5')_{LR}$, where $m_5' = \min\{m_1 m_2, n_1 m_2\}$, $n_5' = \max\{m_1 n_2, n_1 n_2\}$, $\alpha_5' = \min\{m_1 m_2, n_1 m_2\} - \min\{m_1 m_2 - m_1 \alpha_2 - \alpha_1 m_2 + \alpha_1 \alpha_2, n_1 m_2 - n_1 \alpha_2 + \beta_1 m_2 - \beta_1 \alpha_2\}$, $\beta_5' = \max\{m_1 n_2 + m_1 \beta_2 - \alpha_1 n_2 - \alpha_1 \beta_2, n_1 n_2 + n_1 \beta_2 + \beta_1 n_2 + \beta_1 \beta_2\} - \max\{m_1 n_2, n_1 n_2\}$.*

*Proof* Similar to Proposition 4.1.

### 4.2.2  New Product Corresponding to the Existing Product ⊙

In this section, new product corresponding to the existing product ⊙, proposed by Kaur and Kumar [6], is introduced.

**Proposition 4.6** *If $\tilde{A}_1 = (m_1, n_1, \alpha_1, \beta_1)_{LR}$ and $\tilde{A}_2 = (m_2, n_2, \alpha_2, \beta_2)_{LR}$ are two LR flat fuzzy numbers such that $m_1 - \alpha_1 \geq 0$ then $\tilde{A}_1 \odot \tilde{A}_2 = (m_1', n_1', \alpha_1', \beta_1')_{LR}$, where $m_1' = \min\{m_1 m_2, n_1 m_2\}$, $n_1' = \max\{m_1 n_2, n_1 n_2\}$, $\alpha_1' = \min\{m_1 m_2, n_1 m_2\} - \min\{m_1 n_2 + m_1 \beta_2 - \alpha_1 n_2, n_1 m_2 - n_1 \alpha_2 + \beta_1 m_2\}$, $\beta_1' = \max\{m_1 m_2 - m_1 \alpha_2 - \alpha_1 m_2, n_1 n_2 + n_1 \beta_2 + \beta_1 n_2\} - \max\{m_1 n_2, n_1 n_2\}$.*

*Proof* The proposed results may be obtained by considering the following five cases:

**Case (i)**   Neglecting the terms $\alpha_1 \beta_2$ and $\beta_1 \beta_2$ from the results obtained in Case (i) of Proposition 4.1, $A_1^\lambda A_2^\lambda = [m_1 m_2, n_1 n_2]$ for $\lambda = 1$ and $A_1^\lambda A_2^\lambda = [m_1 n_2 + m_1 \beta_2 - \alpha_1 n_2, n_1 n_2 + n_1 \beta_2 + \beta_1 n_2]$ for $\lambda = 0$. Combining the both,
$\tilde{A}_1 \odot \tilde{A}_2 = (m_1'', n_1'', \alpha_1'', \beta_1'')_{LR}$
where $m_1'' = m_1 m_2$, $n_1'' = n_1 n_2$, $\alpha_1'' = m_1 m_2 - m_1 n_2 + m_1 \beta_2 - \alpha_1 n_2$, $\beta_1'' = n_1 n_2 + n_1 \beta_2 + \beta_1 n_2 - n_1 n_2$.

**Case (ii)**  Neglecting the terms $\alpha_1 \beta_2$, $\beta_1 \alpha_2$, $\alpha_1 \alpha_2$ and $\beta_1 \beta_2$ from the results obtained in Case (ii) of Proposition 4.1, $A_1^\lambda A_2^\lambda = [\min\{m_1 n_2 + m_1 \beta_2 - \alpha_1 n_2, n_1 m_2 - n_1 \alpha_2 + \beta_1 m_2\}, \max\{m_1 m_2 - m_1 \alpha_2 - \alpha_1 m_2, n_1 n_2 + n_1 \beta_2 + \beta_1 n_2\}]$ for $\lambda = 0$ and $A_1^\lambda A_2^\lambda = [m_1 m_2, n_1 n_2]$ for $\lambda = 1$. Combining the both,
$\tilde{A}_1 \odot \tilde{A}_2 = (m_2'', n_2'', \alpha_2'', \beta_2'')_{LR}$

where $m_2'' = m_1 m_2, n_2'' = n_1 n_2, \alpha_2'' = m_1 m_2 - \min\{m_1 n_2 + m_1 \beta_2 - \alpha_1 n_2 - m_1 m_2, n_1 m_2 - n_1 \alpha_2 + \beta_1 m_2\}, \beta_2'' = \max\{m_1 m_2 - m_1 \alpha_2 - \alpha_1 m_2, n_1 n_2 + n_1 \beta_2 + \beta_1 n_2\} - n_1 n_2$.

**Case (iii)**  Neglecting the terms $\alpha_1 \beta_2, \beta_1 \alpha_2, \alpha_1 \alpha_2$ and $\beta_1 \beta_2$ from the results obtained in Case (iii) of Proposition 4.1, $A_1^\lambda A_2^\lambda = [n_1 m_2, n_1 n_2]$ for $\lambda = 1$ and $A_1^\lambda A_2^\lambda = [\min\{m_1 n_2 + m_1 \beta_2 - \alpha_1 n_2, n_1 m_2 + \beta_1 m_2 - n_1 \alpha_2\}, \max\{m_1 m_2 - m_1 \alpha_2 - \alpha_1 m_2, n_1 n_2 + n_1 \beta_2 + \beta_1 n_2\}]$ for $\lambda = 0$. Combining the both,
$$\tilde{A}_1 \odot \tilde{A}_2 = (m_3'', n_3'', \alpha_3'', \beta_3'')_{LR}$$
where $m_3'' = n_1 m_2, n_3'' = n_1 n_2, \alpha_3'' = n_1 m_2 - \min\{m_1 n_2 + m_1 \beta_2 - \alpha_1 n_2, n_1 m_2 + \beta_1 m_2 - n_1 \alpha_2\}, \beta_3'' = \max\{m_1 m_2 - m_1 \alpha_2 - \alpha_1 m_2, n_1 n_2 + n_1 \beta_2 + \beta_1 n_2\} - n_1 n_2$.

**Case (iv)**  Neglecting the terms $\alpha_1 \beta_2, \beta_1 \alpha_2, \alpha_1 \alpha_2$ and $\beta_1 \beta_2$ from the results obtained in Case (iv) of Proposition 4.1, $A_1^\lambda A_2^\lambda = [n_1 m_2, m_1 n_2]$ for $\lambda = 1$ and $A_1^\lambda A_2^\lambda = [\min\{m_1 n_2 + m_1 \beta_2 - \alpha_1 n_2, n_1 m_2 - n_1 \alpha_2 + \beta_1 m_2\}, \max\{m_1 m_2 - m_1 \alpha_2 - \alpha_1 m_2, n_1 n_2 + n_1 \beta_2 + \beta_1 n_2\}]$ for $\lambda = 0$. Combining the both,
$$\tilde{A}_1 \odot \tilde{A}_2 = (m_4'', n_4'', \alpha_4'', \beta_4'')_{LR}$$
where $m_4'' = n_1 m_2, n_4'' = m_1 n_2, \alpha_4'' = n_1 m_2 - \min\{m_1 n_2 + m_1 \beta_2 - \alpha_1 n_2, n_1 m_2 - n_1 \alpha_2 + \beta_1 m_2\}, \beta_4'' = \max\{m_1 m_2 - m_1 \alpha_2 - \alpha_1 m_2, n_1 n_2 + n_1 \beta_2 + \beta_1 n_2\} - m_1 n_2$.

**Case (v)**  Neglecting the terms $\beta_1 \alpha_2$ and $\alpha_1 \alpha_2$ from the results obtained in Case (v) of Proposition 4.1, $A_1^\lambda A_2^\lambda = [n_1 m_2, m_1 n_2]$ for $\lambda = 1$ and $A_1^\lambda A_2^\lambda = [n_1 m_2 - n_1 \alpha_2 + \beta_1 m_2, m_1 m_2 - m_1 \alpha_2 - \alpha_1 m_2]$ for $\lambda = 0$. Combining the both,
$$\tilde{A}_1 \odot \tilde{A}_2 = (m_5'', n_5'', \alpha_5'', \beta_5'')_{LR}$$
where $m_5'' = n_1 m_2, n_5'' = m_1 n_2, \alpha_5'' = n_1 m_2 - n_1 m_2 - n_1 \alpha_2 + \beta_1 m_2, \beta_5'' = m_1 m_2 - m_1 \alpha_2 - \alpha_1 m_2 - m_1 n_2$.

Combining the results of all five cases the following result is obtained:
If $\tilde{A}_1 = (m_1, n_1, \alpha_1, \beta_1)_{LR}$ and $\tilde{A}_2 = (m_2, n_2, \alpha_2, \beta_2)_{LR}$ are two *LR* flat fuzzy numbers such that $m_1 - \alpha_1 < 0, m_1 \geq 0$ and $\tilde{A}_2$ is any *LR* flat fuzzy number, then
$$\tilde{A}_1 \odot \tilde{A}_2 = (m_1', n_1', \alpha_1', \beta_1')_{LR}$$
where $m_1' = \min\{m_1 m_2, n_1 m_2\}, n_1' = \max\{m_1 n_2, n_1 n_2\}, \alpha_1' = \min\{m_1 m_2, n_1 m_2\} - \min\{m_1 n_2 + m_1 \beta_2 - \alpha_1 n_2, n_1 m_2 - n_1 \alpha_2 + \beta_1 m_2\}, \beta_1' = \max\{m_1 m_2 - m_1 \alpha_2 - \alpha_1 m_2, n_1 n_2 + n_1 \beta_2 + \beta_1 n_2\} - \max\{m_1 n_2, n_1 n_2\}$.

**Proposition 4.7**  *If* $\tilde{A}_1 = (m_1, n_1, \alpha_1, \beta_1)_{LR}$ *and* $\tilde{A}_2 = (m_2, n_2, \alpha_2, \beta_2)_{LR}$ *are two LR flat fuzzy numbers such that* $m_1 < 0$ *and* $n_1 \geq 0$ *then* $\tilde{A}_1 \odot \tilde{A}_2 = (m_2', n_2', \alpha_2', \beta_2')_{LR}$, *where* $m_2' = \min\{m_1 n_2, n_1 m_2\}, n_2' = \max\{m_1 m_2, n_1 n_2\}, \alpha_2' = \min\{m_1 n_2, n_1 m_2\} - \min\{m_1 n_2 + m_1 \beta_2 - \alpha_1 n_2, n_1 m_2 - n_1 \alpha_2 + \beta_1 m_2\}, \beta_2' = \max\{m_1 m_2 - m_1 \alpha_2 - \alpha_1 m_2, n_1 n_2 + n_1 \beta_2 + \beta_1 n_2\} - \max\{m_1 m_2, n_1 n_2\}$.

*Proof*  Similar to Proposition 4.6.

**Proposition 4.8** *If* $\tilde{A}_1 = (m_1, n_1, \alpha_1, \beta_1)_{LR}$ *and* $\tilde{A}_2 = (m_2, n_2, \alpha_2, \beta_2)_{LR}$ *are two LR flat fuzzy numbers such that* $n_1 < 0$ *and* $n_1 + \beta_1 \geq 0$ *then* $\tilde{A}_1 \odot \tilde{A}_2 = (m'_3, n'_3, \alpha'_3, \beta'_3)_{LR}$, *where* $m'_3 = \min\{m_1 n_2, n_1 n_2\}$, $n'_3 = \max\{n_1 m_2, m_1 m_2\}$, $\alpha'_3 = \min\{m_1 n_2, n_1 n_2\} - \min\{m_1 n_2 + m_1 \beta_2 - \alpha_1 n_2, n_1 m_2 - n_1 \alpha_2 + \beta_1 m_2\}$, $\beta'_3 = \max\{m_1 m_2 - m_1 \alpha_2 - \alpha_1 m_2, n_1 n_2 + n_1 \beta_2 + \beta_1 n_2\} - \max\{n_1 m_2, m_1 m_2\}$.

*Proof* Similar to Proposition 4.6.

**Proposition 4.9** *If* $\tilde{A}_1 = (m_1, n_1, \alpha_1, \beta_1)_{LR}$ *and* $\tilde{A}_2 = (m_2, n_2, \alpha_2, \beta_2)_{LR}$ *are two LR flat fuzzy numbers such that* $n_1 + \beta_1 < 0$ *then* $\tilde{A}_1 \odot \tilde{A}_2 = (m'_4, n'_4, \alpha'_4, \beta'_4)_{LR}$, *where* $m'_4 = \min\{m_1 n_2, n_1 n_2\}$, $n'_4 = \max\{m_1 m_2, n_1 m_2\}$, $\alpha'_4 = \min\{m_1 n_2, n_1 n_2\} - \min\{m_1 n_2 + m_1 \beta_2 - \alpha_1 n_2, n_1 n_2 + n_1 \beta_2 + \beta_1 n_2\}$, $\beta'_4 = \max\{n_1 m_2 - n_1 \alpha_2 + \beta_1 m_2, m_1 m_2 - m_1 \alpha_2 - \alpha_1 m_2\} - \max\{m_1 m_2, n_1 m_2\}$.

*Proof* Similar to Proposition 4.6.

**Proposition 4.10** *If* $\tilde{A}_1 = (m_1, n_1, \alpha_1, \beta_1)_{LR}$ *and* $\tilde{A}_2 = (m_2, n_2, \alpha_2, \beta_2)_{LR}$ *are two LR flat fuzzy numbers such that* $m_1 - \alpha_1 \geq 0$ *then* $\tilde{A}_1 \odot \tilde{A}_2 = (m'_5, n'_5, \alpha'_5, \beta'_5)_{LR}$, *where* $m'_5 = \min\{m_1 m_2, n_1 m_2\}$, $n'_5 = \max\{m_1 n_2, n_1 n_2\}$, $\alpha'_5 = \min\{m_1 m_2, n_1 m_2\} - \min\{m_1 m_2 - m_1 \alpha_2 - \alpha_1 m_2, n_1 m_2 - n_1 \alpha_2 + \beta_1 m_2\}$, $\beta'_5 = \max\{m_1 n_2 + m_1 \beta_2 - \alpha_1 n_2, n_1 n_2 + n_1 \beta_2 + \beta_1 n_2\} - \max\{m_1 n_2, n_1 n_2\}$.

*Proof* Similar to Proposition 4.6.

## 4.3  Limitations of Previous Presented Method

The method, presented in Chap. 3, can be used only for solving such fully fuzzy linear programming problems with equality constraints in which all the parameters are either represented by triangular fuzzy numbers or trapezoidal fuzzy numbers. However, the same method cannot be used for solving fully fuzzy linear programming problems (4.11) and (4.12) in which the parameters are represented by *LR* fuzzy numbers or *LR* flat fuzzy numbers:

$$\text{Maximize/Minimize} \sum_{j=1}^{n} \tilde{c}_j \otimes \tilde{x}_j$$

subject to                                                                                                    (4.11)

$$\sum_{j=1}^{n} \tilde{a}_{ij} \otimes \tilde{x}_j = \tilde{b}_i \ \forall \, i = 1, 2, \ldots, m$$

where $\tilde{a}_{ij} = (a_{ij}, b_{ij}, \alpha_{ij}, \beta_{ij})_{LR}$, $\tilde{x}_j = (x_j, y_j, \alpha''_j, \beta''_j)_{LR}$, $\tilde{b}_i = (b_i, g_i, \gamma_i, \delta_i)_{LR}$ and $\tilde{c}_j = (p_j, q_j, \alpha'_j, \beta'_j)_{LR}$ are *LR* flat fuzzy numbers.

*Example 4.1*

Maximize $((-1, 3, 1, 3)_{LR} \otimes \tilde{x}_1 \oplus (4, 6, 2, 2)_{LR} \otimes \tilde{x}_2)$

subject to

$(2, 5, 5, 2)_{LR} \otimes \tilde{x}_1 \oplus (-1, 5, 1, 2)_{LR} \otimes \tilde{x}_2 = (-17, 45, 25, 46)_{LR}$

$(2, 3, 1, 1)_{LR} \otimes \tilde{x}_1 \oplus (-3, -2, 1, 1)_{LR} \otimes \tilde{x}_2 = (-27, -4, 21, 32)_{LR}$

where $\tilde{x}_1, \tilde{x}_2$ are *LR* flat fuzzy numbers and $L(x) = max\{0, 1 - x\}$, $R(x) = max\{0, 1 - x^2\}$.

$$\text{Maximize/Minimize} \sum_{j=1}^{n} \tilde{c}_j \odot \tilde{x}_j$$

subject to (4.12)

$$\sum_{j=1}^{n} \tilde{a}_{ij} \odot \tilde{x}_j = \tilde{b}_i \ \forall i = 1, 2, \ldots, m$$

where $\tilde{a}_{ij} = (a_{ij}, b_{ij}, \alpha_{ij}, \beta_{ij})_{LR}$, $\tilde{x}_j = (x_j, y_j, \alpha_j'', \beta_j'')_{LR}$, $\tilde{b}_i = (b_i, g_i, \gamma_i, \delta_i)_{LR}$ and $\tilde{c}_j = (p_j, q_j, \alpha_j', \beta_j')_{LR}$ are *LR* flat fuzzy numbers.

*Example 4.2*

Maximize $((1, 2, 2, 2)_{LR} \odot \tilde{x}_1 \oplus (1, 1, 1, 1)_{LR} \odot \tilde{x}_2)$

subject to

$(2, 4, 6, 2)_{LR} \odot \tilde{x}_2 = (2, 8, 10, 4)_{LR}$

$(2, 3, 1, 1)_{LR} \odot \tilde{x}_1 \oplus (-3, -2, 1, 1)_{LR} \odot \tilde{x}_2 = (-2, 7, 6, 7)_{LR}$

where $\tilde{x}_1, \tilde{x}_2$ are *LR* flat fuzzy numbers and $L(x) = max\{0, 1 - x^2\}$, $R(x) = max\{0, 1 - x\}$.

## 4.4 Kaur and Kumar's Method for Solving Fully Fuzzy Linear Programming Problems with Equality Constraints Having *LR* Flat Fuzzy Numbers

In this section, to overcome the limitations of the method, presented in Chap. 3, a method proposed by Kaur and Kumar [6] is presented for solving fully fuzzy linear programming problems (4.11) in which all the parameters are represented as *LR* flat fuzzy numbers. The same method can also be used to find the fuzzy optimal solution of the fully fuzzy linear programming problem (4.11) by replacing $\otimes$ by $\odot$.

The steps of the Kaur and Kumar's method [6] for solving fully fuzzy linear programming problem (4.11) are as follows:

**Step 1** Assuming $\tilde{a}_{ij} = (a_{ij}, b_{ij}, \alpha_{ij}, \beta_{ij})_{LR}$, $\tilde{x}_j = (x_j, y_j, \alpha_j'', \beta_j'')_{LR}$, $\tilde{b}_i = (b_i, g_i, \gamma_i, \delta_i)_{LR}$ and $\tilde{c}_j = (p_j, q_j, \alpha_j', \beta_j')_{LR}$ the fully fuzzy linear programming problem (4.11) can be written as:

$$\text{Maximize/Minimize} \sum_{j=1}^{n}((p_j, q_j, \alpha_j', \beta_j')_{LR} \otimes (x_j, y_j, \alpha_j'', \beta_j'')_{LR})$$

subject to     (4.13)

$$\sum_{j=1}^{n}(a_{ij}, b_{ij}, \alpha_{ij}, \beta_{ij})_{LR} \otimes (x_j, y_j, \alpha_j'', \beta_j'')_{LR} = (b_i, g_i, \gamma_i, \delta_i)_{LR} \ \forall i = 1, 2, \ldots, m$$

where $(x_j, y_j, \alpha_j'', \beta_j'')_{LR}$ is an *LR* flat fuzzy number.

**Step 2** Using the product, presented in Sect. 4.2.1 and assuming $(a_{ij}, b_{ij}, \alpha_{ij}, \beta_{ij})_{LR} \otimes (x_j, y_j, \alpha_j'', \beta_j'')_{LR} = (m_{ij}, n_{ij}, \gamma_{ij}', \delta_{ij}')_{LR}$ the fully fuzzy linear programming problem (4.13) can be written as:

$$\text{Maximize/Minimize} \sum_{j=1}^{n}((p_j, q_j, \alpha_j', \beta_j')_{LR} \otimes (x_j, y_j, \alpha_j'', \beta_j'')_{LR})$$

subject to     (4.14)

$$\sum_{j=1}^{n}(m_{ij}, n_{ij}, \gamma_{ij}', \delta_{ij}')_{LR} = (b_i, g_i, \gamma_i, \delta_i)_{LR} \ \forall i = 1, 2, \ldots, m$$

where $(x_j, y_j, \alpha_j'', \beta_j'')_{LR}$ is an *LR* flat fuzzy number.

**Step 3** Using arithmetic operations, defined in Sect. 4.1.2 and Definition 4.5, the fully fuzzy linear programming problem (4.14) can be written as:

$$\text{Maximize/Minimize} \sum_{j=1}^{n}((p_j, q_j, \alpha_j', \beta_j')_{LR} \otimes (x_j, y_j, \alpha_j'', \beta_j'')_{LR})$$

subject to

$$\sum_{j=1}^{n} m_{ij} = b_i \ \forall i = 1, 2, \ldots, m$$

$$\sum_{j=1}^{n} n_{ij} = g_i \ \forall i = 1, 2, \ldots, m$$     (4.15)

$$\sum_{j=1}^{n} \gamma_{ij}' = \gamma_i \ \forall i = 1, 2, \ldots, m$$

$$\sum_{j=1}^{n} \delta_{ij}' = \delta_i \ \forall i = 1, 2, \ldots, m$$

$$x_j \leq y_j, \alpha_j'' \geq 0, \beta_j'' \geq 0 \ \forall j = 1, 2, \ldots, n$$

**Step 4** Suppose the fuzzy linear programming problem (4.15) have '$l$' basic feasible solutions and $\{x_j^t, y_j^t, \alpha_j''^t, \beta_j''^t\}$ is the $t$th basic feasible solution then our aim is to find that basic feasible solution out of all '$l$' basic feasible solutions corresponding to which the value of objective function is maximum (or minimum) i.e., our aim is to find $\max(\text{or } \min)_{1 \le t \le l}\{\sum_{j=1}^n (p_j, q_j, \alpha_j', \beta_j')_{LR} \otimes (x_j^t, y_j^t, \alpha_j''^t, \beta_j''^t)_{LR}\}$.

Yager [8] has proposed the concept that if $\max(\text{or } \min)_{1 \le t \le l}\{\text{Rank}(\sum_{j=1}^n (p_j, q_j, \alpha_j', \beta_j')_{LR} \otimes (x_j^t, y_j^t, \alpha_j''^t, \beta_j''^t)_{LR})\}$ is $\text{Rank}(\sum_{j=1}^n (p_j, q_j, \alpha_j', \beta_j')_{LR} \otimes (x_j^\theta, y_j^\theta, \alpha_j''^\theta, \beta_j''^\theta)_{LR})$ then $\max(\text{or } \min)_{1 \le t \le l}\{\sum_{j=1}^n (p_j, q_j, \alpha_j', \beta_j')_{LR} \otimes (x_j^t, y_j^t, \alpha_j''^t, \beta_j''^t)_{LR}\}$ will also be $\sum_{j=1}^n (p_j, q_j, \alpha_j', \beta_j')_{LR} \otimes (x_j^\theta, y_j^\theta, \alpha_j''^\theta, \beta_j''^\theta)_{LR})$, i.e. according to the existing method [8] the fuzzy optimal solution of (4.15) can be obtained by solving the problem (4.16):

$$\text{Maximize/Minimize } \Re\left(\sum_{j=1}^n ((p_j, q_j, \alpha_j', \beta_j')_{LR} \otimes (x_j, y_j, \alpha_j'', \beta_j'')_{LR})\right)$$

subject to

$$\sum_{j=1}^n m_{ij} = b_i \ \forall \, i = 1, 2, \ldots, m$$

$$\sum_{j=1}^n n_{ij} = g_i \ \forall \, i = 1, 2, \ldots, m$$

$$\sum_{j=1}^n \gamma_{ij}' = \gamma_i \ \forall \, i = 1, 2, \ldots, m \qquad (4.16)$$

$$\sum_{j=1}^n \delta_{ij}' = \delta_i \ \forall \, i = 1, 2, \ldots, m$$

$$x_j \le y_j, \alpha_j'' \ge 0, \beta_j'' \ge 0 \ \forall \, j = 1, 2, \ldots, n$$

**Step 5** Assuming $(p_j, q_j, \alpha_j', \beta_j')_{LR} \otimes (x_j, y_j, \alpha_j'', \beta_j'')_{LR} = (s_j, t_j, \alpha_j''', \beta_j''')_{LR}$ the problem (4.16) can be written as:

$$\text{Maximize/Minimize } \Re(\sum_{j=1}^{n}(s_j, t_j, \alpha_j''', \beta_j''')_{LR})$$

subject to

$$\sum_{j=1}^{n} m_{ij} = b_i \ \forall \, i = 1, 2, \ldots, m$$

$$\sum_{j=1}^{n} n_{ij} = g_i \ \forall \, i = 1, 2, \ldots, m \qquad (4.17)$$

$$\sum_{j=1}^{n} \gamma_{ij}' = \gamma_i \ \forall \, i = 1, 2, \ldots, m$$

$$\sum_{j=1}^{n} \delta_{ij}' = \delta_i \ \forall \, i = 1, 2, \ldots, m$$

$$x_j \le y_j, \alpha_j'' \ge 0, \beta_j'' \ge 0 \ \forall \, j = 1, 2, \ldots, n$$

**Step 6** Using the linearity property $\Re(\sum_{j=1}^{n} \tilde{A}_i) = \sum_{j=1}^{n} \Re(\tilde{A}_i)$, where $\tilde{A}_i$ is a fuzzy number, problem (4.17) can be converted into (4.18):

$$\text{Maximize/Minimize } \sum_{j=1}^{n} \Re(s_j, t_j, \alpha_j''', \beta_j''')_{LR}$$

subject to

$$\sum_{j=1}^{n} m_{ij} = b_i \ \forall \, i = 1, 2, \ldots, m$$

$$\sum_{j=1}^{n} n_{ij} = g_i \ \forall \, i = 1, 2, \ldots, m \qquad (4.18)$$

$$\sum_{j=1}^{n} \gamma_{ij}' = \gamma_i \ \forall \, i = 1, 2, \ldots, m$$

$$\sum_{j=1}^{n} \delta_{ij}' = \delta_i \ \forall \, i = 1, 2, \ldots, m$$

$$x_j \le y_j, \alpha_j'' \ge 0, \beta_j'' \ge 0 \ \forall \, j = 1, 2, \ldots, n$$

**Step 7** Using the existing formula [8] $\Re(m, n, \alpha, \beta) = \frac{1}{2}(\int_0^1 (m - \alpha L^{-1}(\lambda)) \, d\lambda + \int_0^1 (n + \beta R^{-1}(\lambda)) \, d\lambda)$ problem (4.18) can be converted into (4.19):

$$\text{Maximize/Minimize} \sum_{j=1}^{n} (\frac{1}{2}(\int_{0}^{1} (s_j - \alpha_j''' L^{-1}(\lambda))d\lambda + \int_{0}^{1} (t_j + \beta_j''' R^{-1}(\lambda))d\lambda))$$

subject to

$$\sum_{j=1}^{n} m_{ij} = b_i \ \forall i = 1, 2, \ldots, m$$

$$\sum_{j=1}^{n} n_{ij} = g_i \ \forall i = 1, 2, \ldots, m \qquad (4.19)$$

$$\sum_{j=1}^{n} \gamma_{ij}' = \gamma_i \ \forall i = 1, 2, \ldots, m$$

$$\sum_{j=1}^{n} \delta_{ij}' = \delta_i \ \forall i = 1, 2, \ldots, m$$

$$x_j \leq y_j, \alpha_j'' \geq 0, \beta_j'' \geq 0 \ \forall j = 1, 2, \ldots, n$$

**Step 8** Solve the crisp non-linear programming problem (4.19) by using an appropriate existing method [9] to find the optimal solution $\{x_j^*, y_j^*, \alpha_j''^*, \beta_j''^*\}$.

**Step 9** Find the fuzzy optimal solution $\{\tilde{x}_j^*\}$ of the fully fuzzy linear programming problem (4.11) by putting the values of $x_j^*, y_j^*, \alpha_j''^*$ and $\beta_j''^*$ in $\tilde{x}_j^* = (x_j^*, y_j^*, \alpha_j''^*, \beta_j''^*)$.

**Step 10** Find the fuzzy optimal value of the fully fuzzy linear programming problems (4.11) by putting the values of $\tilde{x}_j^*$, obtained from Step 9, in $\sum_{j=1}^{n} \tilde{c}_j \otimes \tilde{x}_j^*$.

## 4.5 Illustrative Examples

In this section, the method presented in Sect. 4.4 is illustrated with the help of fully fuzzy linear programming problems, chosen in Examples 4.1 and 4.2, which cannot be solved by using the method, presented in Chap. 3.

### 4.5.1 Fuzzy Optimal Solution of the Chosen Fully Fuzzy Linear Programming Problems

In this section, fully fuzzy linear programming problems, chosen in Examples 4.1 and 4.2, are solved by using the method presented in Sect. 4.4.

#### 4.5.1.1    Fuzzy Optimal Solution of the Fully Fuzzy Linear Programming Problem Chosen in Example 4.1

The fuzzy optimal solution of the fully fuzzy linear programming problem, chosen in Example 4.1, can be obtained by using the following steps:

**Step 1** Assuming $\tilde{x}_1 = (x_1, y_1, \alpha_1, \beta_1)_{LR}$ and $\tilde{x}_2 = (x_2, y_2, \alpha_2, \beta_2)_{LR}$ the fully fuzzy linear programming problem, chosen in Example 4.1, can be written as:

Maximize $((-1, 3, 1, 3)_{LR} \otimes (x_1, y_1, \alpha_1, \beta_1)_{LR} \oplus (4, 6, 2, 2)_{LR} \otimes (x_2, y_2, \alpha_2, \beta_2)_{LR})$

subject to

$(2, 5, 5, 2)_{LR} \otimes (x_1, y_1, \alpha_1, \beta_1)_{LR} \oplus (-1, 5, 1, 2)_{LR} \otimes (x_2, y_2, \alpha_2, \beta_2)_{LR} = (-17, 45, 25, 46)_{LR}$

$(2, 3, 1, 1)_{LR} \otimes (x_1, y_1, \alpha_1, \beta_1)_{LR} \oplus (-3, -2, 1, 1)_{LR} \otimes (x_2, y_2, \alpha_2, \beta_2)_{LR} = (-27, -4, 21, 32)_{LR}$

where $(x_1, y_1, \alpha_1, \beta_1)_{LR}, (x_2, y_2, \alpha_2, \beta_2)_{LR}$ are $LR$ flat fuzzy numbers and $L(x) = \max\{0, 1 - x\}, R(x) = \max\{0, 1 - x^2\}$.

**Step 2** Using the product, presented in Sect. 4.2.1, the fully fuzzy linear programming problem, obtained in Step 1, can be written as:

Maximize $((\min\{-y_1, 3x_1\}, \max\{-x_1, 3y_1\}, \min\{-y_1, 3x_1\} - \min\{-2y_1 - 2\beta_1, 6x_1 - 6\alpha_1\}, \max\{-2x_1 + 2\alpha_1, 6y_1 + 6\beta_1\} - \max\{-x_1, 3y_1\})_{LR} \oplus (\min\{4x_2, 6x_2\}, \max\{4y_2, 6y_2\}, \min\{4x_2, 6x_2\} - \min\{2x_2 - 2\alpha_2, 8x_2 - 8\alpha_2\}, \max\{2y_2 + 2\beta_2, 8y_2 + 8\beta_2\} - \max\{4y_2, 6y_2\})_{LR})$

subject to

$(\min\{2x_1, 5x_1\}, \max\{2y_1, 5y_1\}, \min\{2x_1, 5x_1\} - \min\{-3y_1 - 3\beta_1, 7x_1 - 7\alpha_1\}, \max\{7y_1 + 7\beta_1, -3x_1 + 3\alpha_1\} - \max\{2y_1, 5y_1\})_{LR} \oplus (\min\{-y_2, 5x_2\}, \max\{-x_2, 5y_2\}, \min\{-y_2, 5x_2\} - \min\{-2y_2 - 2\beta_2, 7x_2 - 7\alpha_2\}, \max\{-2x_2 + 2\alpha_2, 7y_2 + 7\beta_2\} - \max\{-x_2, 5y_2\})_{LR} = (-17, 45, 25, 46)_{LR}$

$(\min\{2x_1, 3x_1\}, \max\{2y_1, 3y_1\}, \min\{2x_1, 3x_1\} - \min\{x_1 - \alpha_1, 4x_1 - 4\alpha_1\}, \max\{y_1 + \beta_1, 4y_1 + 4\beta_1\} - \max\{2y_1, 3y_1\})_{LR} \oplus (\min\{-3y_2, -2y_2\}, \max\{-3x_2, -2x_2\}, \min\{-3y_2, -2y_2\} - \min\{-4y_2 - 4\beta_2, -y_2 - \beta_2\}, \max\{-4x_2 + 4\alpha_2, -x_2 + \alpha_2\} - \max\{-3x_2, -2x_2\})_{LR} = (-27, -4, 21, 32)_{LR}$

where $(x_1, y_1, \alpha_1, \beta_1)_{LR}$ and $(x_2, y_2, \alpha_2, \beta_2)_{LR}$ are $LR$ flat fuzzy numbers.

**Step 3** Using the arithmetic operations, defined in Sect. 4.1.2 and Definition 4.5, the fully fuzzy linear programming problem, obtained in Step 2, can be written as:

Maximize $(\min\{-y_1, 3x_1\} + \min\{4x_2, 6x_2\}, \max\{-x_1, 3y_1\} + \max\{4y_2, 6y_2\}, \min\{-y_1, 3x_1\} - \min\{-2y_1 - 2\beta_1, 6x_1 - 6\alpha_1\} + \min\{4x_2, 6x_2\} - \min\{2x_2 - 2\alpha_2, 8x_2 - 8\alpha_2\}, \max\{-2x_1 + 2\alpha_1, 6y_1 + 6\beta_1\} - \max\{-x_1, 3y_1\} + \max\{2y_2 + 2\beta_2, 8y_2 + 8\beta_2\} - \max\{4y_2, 6y_2\})_{LR}$

subject to

$$\min\{2x_1, 5x_1\} + \min\{-y_2, 5x_2\} = -17$$
$$\max\{2y_1, 5y_1\} + \max\{-x_2, 5y_2\} = 45$$
$$\min\{2x_1, 3x_1\} + \min\{-3y_2, -2y_2\} = -27$$
$$\max\{2y_1, 3y_1\} + \max\{-3x_2, -2x_2\} = -4$$
$$\min\{2x_1, 3x_1\} - \min\{x_1 - \alpha_1, 4x_1 - 4\alpha_1\}+$$

$$\min\{-3y_2, -2y_2\} - \min\{-4y_2 - 4\beta_2, -y_2 - \beta_2\} = 21$$
$$\max\{y_1 + \beta_1, 4y_1 + 4\beta_1\} - \max\{2y_1, 3y_1\})_{LR} +$$
$$\max\{-4x_2 + 4\alpha_2, -x_2 + \alpha_2\} - \max\{-3x_2, -2x_2\} = 32$$
$$\min\{2x_1, 5x_1\} - \min\{-3y_1 - 3\beta_1, 7x_1 - 7\alpha_1\} +$$
$$\min\{-y_2, 5x_2\} - \min\{-2y_2 - 2\beta_2, 7x_2 - 7\alpha_2\} = 25$$
$$\max\{7y_1 + 7\beta_1, -3x_1 + 3\alpha_1\} - \max\{2y_1, 5y_1\} +$$
$$\max\{-2x_2 + 2\alpha_2, 7y_2 + 7\beta_2\} - \max\{-x_2, 5y_2\})_{LR} = 46$$
$$x_1 \le y_1, \alpha_1 \ge 0, \beta_1 \ge 0, x_2 \le y_2, \alpha_2 \ge 0, \beta_2 \ge 0$$

**Step 4** Using Step 4 of the method, presented in Sect. 4.4, the fuzzy linear programming problem, obtained in Step 3, can be written as:

Maximize $\Re(\min\{-y_1, 3x_1\} + \min\{4x_2, 6x_2\}, \max\{-x_1, 3y_1\} + \max\{4y_2, 6y_2\}, \min\{-y_1, 3x_1\} - \min\{-2y_1 - 2\beta_1, 6x_1 - 6\alpha_1\} + \min\{4x_2, 6x_2\} - \min\{2x_2 - 2\alpha_2, 8x_2 - 8\alpha_2\}, \max\{-2x_1 + 2\alpha_1, 6y_1 + 6\beta_1\} - \max\{-x_1, 3y_1\} + \max\{2y_2 + 2\beta_2, 8y_2 + 8\beta_2\} - \max\{4y_2, 6y_2\})_{LR}$

subject to

$$\min\{2x_1, 5x_1\} + \min\{-y_2, 5x_2\} = -17$$
$$\max\{2y_1, 5y_1\} + \max\{-x_2, 5y_2\} = 45$$
$$\min\{2x_1, 3x_1\} + \min\{-3y_2, -2y_2\} = -27$$
$$\max\{2y_1, 3y_1\} + \max\{-3x_2, -2x_2\} = -4$$
$$\min\{2x_1, 3x_1\} - \min\{x_1 - \alpha_1, 4x_1 - 4\alpha_1\} +$$
$$\min\{-3y_2, -2y_2\} - \min\{-4y_2 - 4\beta_2, -y_2 - \beta_2\} = 21$$
$$\max\{y_1 + \beta_1, 4y_1 + 4\beta_1\} - \max\{2y_1, 3y_1\})_{LR} +$$
$$\max\{-4x_2 + 4\alpha_2, -x_2 + \alpha_2\} - \max\{-3x_2, -2x_2\} = 32$$
$$\min\{2x_1, 5x_1\} - \min\{-3y_1 - 3\beta_1, 7x_1 - 7\alpha_1\} +$$
$$\min\{-y_2, 5x_2\} - \min\{-2y_2 - 2\beta_2, 7x_2 - 7\alpha_2\} = 25$$
$$\max\{7y_1 + 7\beta_1, -3x_1 + 3\alpha_1\} - \max\{2y_1, 5y_1\} +$$
$$\max\{-2x_2 + 2\alpha_2, 7y_2 + 7\beta_2\} - \max\{-x_2, 5y_2\})_{LR} = 46$$
$$x_1 \le y_1, \alpha_1 \ge 0, \beta_1 \ge 0, x_2 \le y_2, \alpha_2 \ge 0, \beta_2 \ge 0$$

**Step 5** Using Step 6 and Step 7 of the method, presented in Sect. 4.4, the fuzzy linear programming problem, obtained in Step 4, can be written as:

Maximize $(\frac{17}{24}x_1 - \frac{5}{12}\alpha_1 + \frac{5}{2}x_2 - \frac{5}{4}\alpha_2 + \frac{7}{8}y_1 + \frac{3}{4}\beta_1 + \frac{5}{2}y_2 + \frac{5}{3}\beta_2 - \frac{1}{8}|y_1 + 3x_1| + \frac{1}{12}|x_1 + 3y_1| + \frac{1}{3}|3y_1 + 3\beta_1 + x_1 - \alpha_1| - \frac{1}{4}|y_1 + \beta_1 + 3x_1 - 3\alpha_1| - \frac{1}{4}|x_2| + \frac{1}{6}|y_2| + |y_2 + \beta_2| - \frac{3}{4}|x_2 - \alpha_2|)$

subject to

$$\frac{7}{2}x_1 - \frac{3}{2}|x_1| - \frac{1}{2}y_2 + \frac{5}{2}x_2 - \frac{1}{2}|y_2 + 5x_2| = -17$$
$$\frac{7}{2}y_1 + \frac{3}{2}|y_1| - \frac{1}{2}x_2 + \frac{5}{2}y_2 + \frac{1}{2}|x_2 + 5y_2| = 45$$
$$\frac{5}{2}x_1 - \frac{1}{2}|x_1| - \frac{5}{2}y_2 - \frac{1}{2}|y_2| = -27$$
$$\frac{5}{2}y_1 + \frac{1}{2}|y_1| - \frac{5}{2}x_2 + \frac{1}{2}|x_2| = -4$$
$$-\frac{1}{2}|x_1| - \frac{1}{2}|y_2| + \frac{5}{2}\alpha_1 + \frac{3}{2}|x_1 - \alpha_1| + \frac{5}{2}\beta_2 + \frac{3}{2}|y_2 + \beta_2| = 21$$
$$\frac{5}{2}\beta_1 + \frac{3}{2}|y_1 + \beta_1| + \frac{5}{2}\alpha_2 + \frac{3}{2}|x_2 - \alpha_2| - \frac{1}{2}|y_1| - \frac{1}{2}|x_2| = 32$$
$$-\frac{3}{2}|x_1| + \frac{1}{2}y_2 - x_2 - \frac{1}{2}|5x_2 + y_2| + \frac{3}{2}y_1 + \frac{3}{2}\beta_1 + \frac{7}{2}\alpha_1 +$$

$$\tfrac{1}{2}|7x_1+3y_1-7\alpha_1+3\beta_1|+\tfrac{7}{2}\alpha_2+\beta_2+|\tfrac{7}{2}x_2+y_2-\tfrac{7}{2}\alpha_2+\beta_2| = 25$$
$$\tfrac{7}{2}\beta_1-\tfrac{3}{2}x_1+\tfrac{3}{2}\alpha_1+\tfrac{1}{2}|3x_1+7y_1-3\alpha_1+7\beta_1|-\tfrac{1}{2}x_2+\alpha_2+y_2+$$
$$\tfrac{7}{2}\beta_2+|x_2+\tfrac{7}{2}y_2-\alpha_2+\tfrac{7}{2}\beta_2|-\tfrac{3}{2}|y_1|-\tfrac{1}{2}|x_2+5y_2| = 46$$
$$x_1 \le y_1, \alpha_1 \ge 0, \beta_1 \ge 0, x_2 \le y_2, \alpha_2 \ge 0, \beta_2 \ge 0$$

**Step 6** The optimal solution of the crisp non-linear programming problem, obtained in Step 5, is $x_1 = -2, y_1 = 2, \alpha_1 = 0, \beta_1 = 1, x_2 = 5, y_2 = 7, \alpha_2 = 9$ and $\beta_2 = 3$.

**Step 7** Putting the values of $x_1, y_1, \alpha_1, \beta_1, x_2, y_2, \alpha_2$ and $\beta_2$ in $\tilde{x}_1 = (x_1, y_1, \alpha_1, \beta_1)$ and $\tilde{x}_2 = (x_2, y_2, \alpha_2, \beta_2)$, the exact fuzzy optimal solution is $\tilde{x}_1 = (-2, 2, 0, 1)_{LR}$, $\tilde{x}_2 = (5, 7, 9, 3)_{LR}$.

**Step 8** Putting the values of $\tilde{x}_1$ and $\tilde{x}_2$, obtained from Step 7, in the objective function the fuzzy optimal value is $(14, 48, 58, 50)_{LR}$.

### 4.5.1.2　Fuzzy Optimal Solution of the Fully Fuzzy Linear Programming Problem Chosen in Example 4.2

The fuzzy optimal solution of the fully fuzzy linear programming problem, chosen in Example 4.2, can be obtained by using the following steps:

**Step 1** Assuming $\tilde{x}_1 = (x_1, y_1, \alpha_1, \beta_1)_{LR}$ and $\tilde{x}_2 = (x_2, y_2, \alpha_2, \beta_2)_{LR}$ the fully fuzzy linear programming problem, chosen in Example 4.1, can be written as:
Maximize $((1, 2, 2, 2)_{LR} \odot (x_1, y_1, \alpha_1, \beta_1) \oplus (1, 1, 1, 1)_{LR} \odot (x_2, y_2, \alpha_2, \beta_2))$

subject to
$(2, 4, 6, 2)_{LR} \odot (x_2, y_2, \alpha_2, \beta_2) = (2, 8, 10, 4)_{LR}$
$(2, 3, 1, 1)_{LR} \odot (x_1, y_1, \alpha_1, \beta_1) \oplus (-3, -2, 1, 1)_{LR} \odot (x_2, y_2, \alpha_2, \beta_2) = (-2, 7, 6, 7)_{LR}$
where $(x_1, y_1, \alpha_1, \beta_1)_{LR}, (x_2, y_2, \alpha_2, \beta_2)_{LR}$ are $LR$ flat fuzzy numbers and $L(x) = \max\{0, 1 - x^2\}, R(x) = \max\{0, 1 - x\}$.

**Step 2** Using the product, presented in Sect. 4.2.2, the fully fuzzy linear programming problem, obtained in Step 1, can be written as:
Maximize $((\min\{x_1, 2x_1\}, \max\{y_1, 2y_1\}, \min\{x_1, 2x_1\} - \min\{-y_1 + \beta_1, 4x_1 - 2\alpha_1\},$
$\max\{-x_1 - \alpha_1, 4y_1 + 2\beta_1\} - \max\{y_1, 2y_1\})_{LR} \oplus (\min\{x_2, x_2\}, \max\{y_2, y_2\}, \min\{x_2, x_2\}$
$-\min\{-\alpha_2, 2x_2 - \alpha_2\}, \max\{\beta_2, 2y_2 + \beta_2\} - \max\{y_2, y_2\})_{LR})$

subject to

$(\min\{2x_2, 4x_2\}, \max\{2y_2, 4y_2\}, \min\{2x_2, 4x_2\} - \min\{-4y_2 + 2\beta_2, 6x_2 - 4\alpha_2\},$
$\max\{-4x_2 - 2\alpha_2, 6y_2 + 4\beta_2\} - \max\{2y_2, 4y_2\})_{LR} = (2, 8, 10, 4)_{LR}$
$(\min\{2x_1, 3x_1\}, \max\{2y_1, 3y_1\}, \min\{2x_1, 3x_1\} - \min\{x_1 - 2\alpha_1, 4x_1 - 3\alpha_1\}, \max\{y_1 +$
$2\beta_1, 4y_1 + 3\beta_1\} - \max\{2y_1, 3y_1\})_{LR} \oplus (\min\{-3y_2, -2y_2\}, \max\{-3x_2, -2x_2\},$
$\min\{-3y_2, -2y_2\} - \min\{-4y_2 - 3\beta_2, -y_2 - 2\beta_2\}, \max\{-4x_2 + 3\alpha_2, -x_2 + 2\alpha_2\} -$
$\max\{-3x_2, -2x_2\})_{LR} = (-2, 7, 6, 7)_{LR}$
where $(x_1, y_1, \alpha_1, \beta_1)_{LR}$ and $(x_2, y_2, \alpha_2, \beta_2)_{LR}$ are $LR$ flat fuzzy numbers.

**Step 3** Using the arithmetic operations, defined in Sect. 4.1.2 and Definition 4.5, the fully fuzzy linear programming problem, obtained in Step 2, can be written as:

Maximize $(\min\{x_1, 2x_1\} + \min\{x_2, x_2\}, \max\{y_1, 2y_1\} + \max\{y_2, y_2\}, \min\{x_1, 2x_1\} - \min\{-y_1 + \beta_1, 4x_1 - 2\alpha_1\} + \min\{x_2, x_2\} - \min\{-\alpha_2, 2x_2 - \alpha_2\}, \max\{-x_1 - \alpha_1, 4y_1 + 2\beta_1\} - \max\{y_1, 2y_1\} + \max\{\beta_2, 2y_2 + \beta_2\} - \max\{y_2, y_2\})_{LR}$
subject to

$$\min\{2x_2, 4x_2\} = 2$$
$$\max\{2y_2, 4y_2\} = 8$$
$$\min\{2x_2, 4x_2\} - \min\{-4y_2 + 2\beta_2, 6x_2 - 4\alpha_2\} = 10$$
$$\max\{-4x_2 - 2\alpha_2, 6y_2 + 4\beta_2\} - \max\{2y_2, 4y_2\} = 4$$
$$\min\{2x_1, 3x_1\} + \min\{-3y_2, -2y_2\} = -2$$
$$\max\{2y_1, 3y_1\} + \max\{-3x_2, -2x_2\} = 7$$
$$\min\{2x_1, 3x_1\} - \min\{x_1 - 2\alpha_1, 4x_1 - 3\alpha_1\} +$$
$$\min\{-3y_2, -2y_2\} - \min\{-4y_2 - 3\beta_2, -y_2 - 2\beta_2\} = 6$$
$$\max\{y_1 + 2\beta_1, 4y_1 + 3\beta_1\} - \max\{2y_1, 3y_1\} +$$
$$\max\{-4x_2 + 3\alpha_2, -x_2 + 2\alpha_2\} - \max\{-3x_2, -2x_2\} = 7$$
$$x_1 \le y_1, \alpha_1 \ge 0, \beta_1 \ge 0, x_2 \le y_2, \alpha_2 \ge 0, \beta_2 \ge 0$$

**Step 4** Using Step 4 of the method, presented in Sect. 4.4, the fuzzy linear programming problem, obtained in Step 3, can be written as:
Maximize $\Re(\min\{x_1, 2x_1\} + \min\{x_2, x_2\}, \max\{y_1, 2y_1\} + \max\{y_2, y_2\}, \min\{x_1, 2x_1\} - \min\{-y_1 + \beta_1, 4x_1 - 2\alpha_1\} + \min\{x_2, x_2\} - \min\{-\alpha_2, 2x_2 - \alpha_2\}, \max\{-x_1 - \alpha_1, 4y_1 + 2\beta_1\} - \max\{y_1, 2y_1\} + \max\{\beta_2, 2y_2 + \beta_2\} - \max\{y_2, y_2\})_{LR}$
subject to

$$\min\{2x_2, 4x_2\} = 2$$
$$\max\{2y_2, 4y_2\} = 8$$
$$\min\{2x_2, 4x_2\} - \min\{-4y_2 + 2\beta_2, 6x_2 - 4\alpha_2\} = 10$$
$$\max\{-4x_2 - 2\alpha_2, 6y_2 + 4\beta_2\} - \max\{2y_2, 4y_2\} = 4$$
$$\min\{2x_1, 3x_1\} + \min\{-3y_2, -2y_2\} = -2$$
$$\max\{2y_1, 3y_1\} + \max\{-3x_2, -2x_2\} = 7$$
$$\min\{2x_1, 3x_1\} - \min\{x_1 - 2\alpha_1, 4x_1 - 3\alpha_1\} +$$
$$\min\{-3y_2, -2y_2\} - \min\{-4y_2 - 3\beta_2, -y_2 - 2\beta_2\} = 6$$
$$\max\{y_1 + 2\beta_1, 4y_1 + 3\beta_1\} - \max\{2y_1, 3y_1\} +$$
$$\max\{-4x_2 + 3\alpha_2, -x_2 + 2\alpha_2\} - \max\{-3x_2, -2x_2\} = 7$$
$$x_1 \le y_1, \alpha_1 \ge 0, \beta_1 \ge 0, x_2 \le y_2, \alpha_2 \ge 0, \beta_2 \ge 0$$

**Step 5** Using Step 6 and Step 7 of the method, presented in Sect. 4.4, the fuzzy linear programming problem, obtained in Step 4, can be written as:
Maximize $(\frac{19}{24}x_1 + \frac{17}{24}y_1 - \frac{11}{24}\alpha_1 + \frac{5}{12}\beta_1 + \frac{1}{2}x_2 + \frac{1}{2}y_2 - \frac{1}{3}\alpha_2 + \frac{1}{4}\beta_2 - \frac{1}{12}|x_1| + \frac{1}{8}|y_1| - \frac{1}{3}|x_2| + \frac{1}{4}|y_2| + \frac{1}{4}|2y_1 + \beta_1 + \frac{1}{2}x_1 + \frac{1}{2}\alpha_1| - \frac{1}{3}|2x_1 - \alpha_1 + \frac{1}{2}y_1 - \frac{1}{2}\beta_1|)$
subject to

$$3x_2 - |x_2| = 2$$
$$3y_2 + |y_2| = 8$$
$$-|x_2| + 2y_2 - \beta_2 + 2\alpha_2 + |2y_2 - \beta_2 + 3x_2 - 2\alpha_2| = 10$$
$$-2x_2 - \alpha_2 + 2\beta_2 + |3y_2 + 2\beta_2 + 2x_2 + \alpha_2| - |y_2| = 4$$
$$\frac{5}{2}x_1 - \frac{1}{2}|x_1| - \frac{5}{2}y_2 - \frac{1}{2}|y_2| = -2$$
$$\frac{5}{2}y_1 + \frac{1}{2}|y_1| - \frac{5}{2}x_2 + \frac{1}{2}|x_2| = 7$$
$$\frac{5}{2}\alpha_1 - \frac{1}{2}|x_1| + \frac{1}{2}|3x_1 - \alpha_1| - \frac{1}{2}|y_2| + \frac{5}{2}\beta_2 + \frac{1}{2}|3y_2 + \beta_2| = 6$$

$$\tfrac{5}{2}\beta_1 + \tfrac{1}{2}|3y_1 + \beta_1| - \tfrac{1}{2}|y_1| + \tfrac{5}{2}\alpha_2 + \tfrac{1}{2}|3x_2 - \alpha_2| - \tfrac{1}{2}|x_2| = 7$$
$$x_1 \le y_1, \alpha_1 \ge 0, \beta_1 \ge 0, x_2 \le y_2, \alpha_2 \ge 0, \beta_2 \ge 0$$

**Step 6** The optimal solution of the crisp non-linear programming problem, obtained in Step 5, is $x_1 = 2, y_1 = 3, \alpha_1 = 1, \beta_1 = 1, x_2 = 1, y_2 = 2, \alpha_2 = 0$ and $\beta_2 = 0$.

**Step 7** Putting the values of $x_1, y_1, \alpha_1, \beta_1, x_2, y_2, \alpha_2$ and $\beta_2$ in $\tilde{x}_1 = (x_1, y_1, \alpha_1, \beta_1)$ and $\tilde{x}_2 = (x_2, y_2, \alpha_2, \beta_2)$, the exact fuzzy optimal solution is $\tilde{x}_1 = (2, 3, 1, 1)_{LR}, \tilde{x}_2 = (1, 2, 0, 0)_{LR}$.

**Step 8** Putting the values of $\tilde{x}_1$ and $\tilde{x}_2$, obtained from Step 7, in the objective function the fuzzy optimal value is $(3, 8, 5, 10)_{LR}$.

## 4.6 Advantages of the Kaur and Kumar's Method

The main advantage of the Kaur and Kumar's method [6], presented in Sect. 4.4, is that all the fully fuzzy linear programming problems which can be solved by using the methods, presented in previous chapters, can also be solved by using the method presented in this chapter. However, there exist several fully fuzzy linear programming problems which cannot be solved by using the methods, presented in previous chapters, but can be solved by using the method, presented in this chapter.

## 4.7 Comparative Study

The results of the chosen fully fuzzy linear programming problems, obtained by using the method, presente in Chap. 3 and the method presented in this chapter, are shown in Table 4.1.

**Table 4.1** Results of the chosen fully fuzzy linear programming problems

| Example | Fuzzy optimal value | |
|---------|---------------------|---|
| | Method presented in Chap. 3 | Method presented in this chapter |
| 3.1 | $(-10, 2, \tfrac{831}{25}, 68)$ | $(2, \tfrac{831}{25}, 12, \tfrac{869}{25})_{LR}$ |
| 3.2 | $(-44, 14, 48, 98)$ | $(14, 48, 58, 50)_{LR}$ |
| 4.1 | Not applicable | $(14, 48, 58, 50)_{LR}$ |
| 4.2 | Not applicable | $(3, 8, 5, 10)_{LR}$ |

The results, presented in Table 4.1 can be explained as follows:

(i)  In the problems, chosen in Examples 3.1 and 3.2, all the parameters are represented by trapezoidal fuzzy numbers while in the problems, chosen in Examples 4.1 and 4.2, all the parameters are represented by unrestricted $LR$ flat fuzzy numbers. So, as discussed in Sect. 4.3 the problems, chosen in Examples 3.1 and 3.2, can be solved by using the method, presented in Chap. 3. However, the problems, chosen in Examples 4.1 and 4.2, cannot be solved by using the method, presented in Chap. 3.

(ii)  Since, the method presented in this chapter can be used to find the fuzzy optimal solution of fully fuzzy linear programming problems with unrestricted $LR$ flat fuzzy parameters. So, all the chosen problems can be solved by using the method presented in this chapter.

## 4.8  Conclusions

On the basis of present study, it can be concluded that the method, presented in this chapter, can be used to solve all the fully fuzzy linear programming problems with equality constraints. Hence, it is better to use the method, presented in this chapter, as compared to the methods, presented in previous chapters for solving fully fuzzy linear programming problems with equality constraints.

## References

1. Dubois, D., Prade, H.: Fuzzy Sets and Systems, Theory and Applications. Academic Press, New York (1980)
2. Ramik, J., Rimanek, J.: Inequality relation between fuzzy numbers and its use in fuzzy optimization. Fuzzy Sets Syst. **16**, 123–138 (1985)
3. Rommelfanger, H.: Interactive decision making in fuzzy linear optimization problems. Eur. J. Oper. Res. **41**, 210–217 (1989)
4. Heilpern, S.: Representation and application of fuzzy numbers. Fuzzy Sets Syst. **91**, 259–268 (1997)
5. Liang, T.F.: Project management decisions using fuzzy linear programming. Int. J. Syst. Sci. **37**, 1141–1152 (2006)
6. Kaur, J., Kumar, A.: Mehar's method for solving fully fuzzy linear programming problems with $LR$ fuzzy parameters. Appl. Math. Model. **37**, 7142–7153 (2013)
7. Dehghan, M., Hashemi, B., Ghatee, M.: Computational methods for solving fully fuzzy linear systems. Appl. Math. Comput. **179**, 328–343 (2006)
8. Yager, R.R.: A procedure for ordering fuzzy subsets of the unit interval. Inf. Sci. **24**, 143–161 (1981)
9. Taha, H.A.: Operations Research: An Introduction. Prentice-Hall, New Jersey (2003)

# Chapter 5
# Fuzzy Optimal Solution of Fully Fuzzy Linear Programming Problems with Inequality Constraints Having LR Flat Fuzzy Numbers

In this chapter, limitations of the existing methods for solving fuzzy linear programming problems and fully fuzzy linear programming problems with inequality constraints are pointed out. To overcome the limitations of the existing methods, two new methods proposed by Kumar and Kaur [1] for solving fully fuzzy linear programming problems with inequality constraints, are presented. The advantages of the methods, presented in this chapter, over the existing methods are also discussed.

## 5.1 Existing Method for Solving Fully Fuzzy Linear Programming Problems with Inequality Constraints

Allahviranloo et al. [2] proposed the following method for solving fully fuzzy linear programming problems (5.1):

$$\text{Maximize/Minimize} \sum_{j=1}^{n} \tilde{c}_j \odot \tilde{x}_j$$

subject to $\hspace{6cm}$ (5.1)

$$\sum_{j=1}^{n} \tilde{a}_{ij} \odot \tilde{x}_j \preceq, \approx, \succeq \tilde{b}_i \ \forall \ i = 1, 2, \ldots, m$$

where $\tilde{x}_j$ is a non-negative $LR$ fuzzy number and $\tilde{a}_{ij}$, $\tilde{b}_i$, $\tilde{c}_j$ are non-negative or non-positive $LR$ fuzzy numbers.

---

The contents of this chapter are published in *Journal of Intelligent and Fuzzy Systems* 26 (2014) 337–344.

© Springer International Publishing Switzerland 2016
J. Kaur and A. Kumar, *An Introduction to Fuzzy Linear Programming Problems*, Studies in Fuzziness and Soft Computing 340,
DOI 10.1007/978-3-319-31274-3_5

**Step 1** Assuming $N_1 = \{j \ : \ 1 \leq j \leq n \text{ and } \tilde{c}_j$ is non-negative $LR$ fuzzy number$\}$, $N_2 = \{j \ : \ 1 \leq j \leq n \text{ and } \tilde{c}_j$ is non-positive $LR$ fuzzy number$\}$, $N_3 = \{j \ : \ 1 \leq j \leq n \text{ and } \tilde{a}_{ij}$ is non-negative $LR$ fuzzy number$\}$, $N_4 = \{j \ : \ 1 \leq j \leq n \text{ and } \tilde{a}_{ij}$ is non-positive $LR$ fuzzy number$\}$ where $N_1 \cup N_2 = N, N_3 \cup N_4 = N, N_1 \cap N_2 = \phi, N_3 \cap N_4 = \phi$, the fully fuzzy linear programming problem (5.1) can be written as:

$$\text{Maximize/Minimize} \sum_{j:j\in N_1} \tilde{c}_j \odot \tilde{x}_j \oplus \sum_{j:j\in N_2} \tilde{c}_j \odot \tilde{x}_j$$

subject to                                                                                          (5.2)

$$\sum_{j:j\in N_3} \tilde{a}_{ij} \odot \tilde{x}_j \oplus \sum_{j:j\in N_4} \tilde{a}_{ij} \odot \tilde{x}_j \preceq, \approx, \succeq \tilde{b}_i \ \forall i = 1, 2, \ldots, m$$

where $\tilde{x}_j$ is a non-negative $LR$ fuzzy number.

**Step 2** Assuming $\tilde{c}_j = (p_j, \alpha'_j, \beta'_j)_{LR}, \tilde{a}_{ij} = (a_{ij}, \alpha_{ij}, \beta_{ij})_{LR}, \tilde{x}_j = (x_j, \alpha''_j, \beta''_j)_{LR}$ and $\tilde{b}_i = (b_i, \gamma_i, \delta_i)_{LR}$ the fully fuzzy linear programming problem (5.2) can be written as:

$$\text{Maximize/Minimize} \sum_{j\in N_1} (p_j, \alpha'_j, \beta'_j)_{LR} \odot (x_j, \alpha''_j, \beta''_j)_{LR} \oplus \sum_{j\in N_2} (p_j, \alpha'_j, \beta'_j)_{LR} \odot (x_j, \alpha''_j, \beta''_j)_{LR}$$

subject to                                                                                          (5.3)

$$\sum_{j\in N_3} (a_{ij}, \alpha_{ij}, \beta_{ij})_{LR} \odot (x_j, \alpha''_j, \beta''_j)_{LR} \oplus \sum_{j\in N_4} (a_{ij}, \alpha_{ij}, \beta_{ij})_{LR} \odot (x_j, \alpha''_j, \beta''_j)_{LR} \preceq, \approx, \succeq$$

$$(b_i, \gamma_i, \delta_i)_{LR} \ \forall i = 1, 2, \ldots, m$$

where $\tilde{x}_j$ is a non-negative $LR$ fuzzy number.

**Step 3** Using the arithmetic operations, presented in Sect. 4.1.2, the fully fuzzy linear programming problem (5.3) can be written as:

$$\text{Maximize/Minimize} \sum_{j:j\in N_1} (p_jx_j, p_j\alpha''_j + \alpha'_jx_j, p_j\beta''_j + \beta'_jx_j)_{LR} \oplus \sum_{j:j\in N_2} (p_jx_j, \alpha'_jx_j -$$

$$p_j\alpha''_j, \beta'_jx_j - p_j\beta''_j)_{LR}$$

subject to                                                                                          (5.4)

$$\sum_{j:j\in N_3} (a_{ij}x_j, a_{ij}\alpha''_j + \alpha_{ij}x_j, a_{ij}\beta''_j + \beta_{ij}x_j)_{LR} \oplus \sum_{j:j\in N_4} (a_{ij}x_j, \alpha_{ij}x_j - a_{ij}\alpha''_j, \beta_{ij}x_j -$$

$$a_{ij}\beta''_j)_{LR} \preceq, \approx, \succeq (b_i, \gamma_i, \delta_i)_{LR} \ \forall i = 1, 2, \ldots, m$$

where $\tilde{x}_j$ is a non-negative $LR$ fuzzy number.

**Step 4** Using the existing ranking approach [3], the fully fuzzy linear programming problem (5.4) can be written as:

$$\text{Maximize/Minimize } \Re(\sum_{j:j\in N_1}(p_jx_j, p_j\alpha''_j + \alpha'_jx_j, p_j\beta''_j + \beta'_jx_j)_{LR}$$

$$\oplus \sum_{j:j\in N_2}(p_jx_j, \alpha'_jx_j - p_j\alpha''_j, \beta'_jx_j - p_j\beta''_j)_{LR})$$

subject to $\qquad\qquad\qquad\qquad\qquad\qquad\qquad\qquad\qquad$ (5.5)

$$\Re(\sum_{j:j\in N_3}(a_{ij}x_j, a_{ij}\alpha''_j + \alpha_{ij}x_j, a_{ij}\beta''_j + \beta_{ij}x_j)_{LR} \oplus \sum_{j:j\in N_4}(a_{ij}x_j, \alpha_{ij}x_j - a_{ij}\alpha''_j, \beta_{ij}x_j -$$

$$a_{ij}\beta''_j)_{LR}) \leq, =, \geq \Re(b_i, \gamma_i, \delta_i)_{LR} \ \forall i = 1, 2, \ldots, m$$

$$x_j - \alpha''_j \geq 0, \alpha''_j \geq 0, \beta''_j \geq 0 \ \forall j \in N$$

**Step 5** Using the linearity property $\Re(\sum_{j=1}^{n} \tilde{A}_i) = \sum_{j=1}^{n} \Re(\tilde{A}_i)$, where $\tilde{A}_i$ is a fuzzy number, the fully fuzzy linear programming problem (5.5) can be written as:

$$\text{Maximize/Minimize } \sum_{j:j\in N_1} \Re(p_jx_j, p_j\alpha''_j + \alpha'_jx_j, p_j\beta''_j + \beta'_jx_j)_{LR}$$

$$+ \sum_{j:j\in N_2} \Re(p_jx_j, \alpha'_jx_j - p_j\alpha''_j, \beta'_jx_j - p_j\beta''_j)_{LR}$$

subject to

$$\sum_{j:j\in N_3} \Re(a_{ij}x_j, a_{ij}\alpha''_j + \alpha_{ij}x_j, a_{ij}\beta''_j + \beta_{ij}x_j)_{LR} \qquad\qquad (5.6)$$

$$+ \sum_{j:j\in N_4} \Re(a_{ij}x_j, \alpha_{ij}x_j - a_{ij}\alpha''_j, \beta_{ij}x_j - a_{ij}\beta''_j)_{LR}$$

$$\leq, =, \geq \Re(b_i, \gamma_i, \delta_i)_{LR} \ \forall i = 1, 2, \ldots, m$$

$$x_j - \alpha''_j \geq 0, \alpha''_j \geq 0, \beta''_j \geq 0 \ \forall j \in N$$

**Step 6** Solve the crisp linear programming problem (5.6) to find the optimal solution $\{x_j^*, \alpha''_j{}^*, \beta''_j{}^*\}$.

**Step 7** Find the fuzzy optimal solution $\{\tilde{x}_j^*\}$ of the fully fuzzy linear programming problem (5.1) by putting the values of $x_j^*$, $\alpha''_j{}^*$ and $\beta''_j{}^*$ in $\tilde{x}_j^* = (x_j^*, \alpha''_j{}^*, \beta''_j{}^*)$.

**Step 8** Find the fuzzy optimal value by putting the values of $\tilde{x}_j^*$, obtained from Step 7, in $\sum_{j=1}^{n} \tilde{c}_j \odot \tilde{x}_j^*$.

## 5.2 Applicability of the Existing Methods

In this section, different types of fuzzy linear programming problems and fully fuzzy linear programming problems, which can be solved by using the existing methods [2, 4–8], are discussed:

(i) The existing methods [5, 7] can be used for solving the following type of fuzzy linear programming problems:

$$\text{Maximize/Minimize} \sum_{j=1}^{n} \tilde{c}_j x_j$$

subject to

$$\sum_{j=1}^{n} \tilde{a}_{ij} x_j \preceq, \approx, \succeq \tilde{b}_i \ \forall i = 1, 2, \ldots, m \tag{5.7}$$

$$x_j \geq 0 \ \forall j = 1, 2, \ldots, n$$

where $\tilde{a}_{ij}$, $\tilde{b}_i$ and $\tilde{c}_j$ are trapezoidal fuzzy numbers.

*Example 5.1* [5, 7]

$$\text{Maximize} ((2, 3, 1, 1)x_1 \oplus (3, 4, 1, 2)x_2)$$

subject to

$$(1, 2, 1, 1)x_1 \oplus (2, 3, 1, 2)x_2 \preceq (5, 6, 2, 2)$$

$$(2, 3, 1, 3)x_1 \oplus (1, 2, 1, 1)x_2 \preceq (4, 6, 2, 1)$$

$$x_1, x_2 \geq 0$$

(ii) Nasseri et al. [8] proposed a method for solving the following type of fuzzy linear programming problems:

$$\text{Maximize/Minimize} \sum_{j=1}^{n} \tilde{c}_j x_j$$

subject to

$$\sum_{j=1}^{n} a_{ij} x_j \leq, =, \geq b_i \ \forall i = 1, 2, \ldots, m \tag{5.8}$$

$$x_j \geq 0 \ \forall j = 1, 2, \ldots, n$$

where $a_{ij}$, $b_i$ are real numbers and $\tilde{c}_j$ is a trapezoidal fuzzy number.

*Example 5.2* [8]

$$\text{Maximize} ((5, 8, 2, 5)x_1 \oplus (6, 10, 2, 6)x_2)$$

subject to

$$2x_1 + 3x_2 \leq 6$$

$$5x_1 + 4x_2 \leq 10$$

$$x_1, x_2 \geq 0$$

(iii) In the existing methods [5, 7, 8] either the cost coefficients are assumed as crisp numbers or the decision variables are assumed as crisp numbers i.e., in none of the existing methods [5, 7, 8] cost coefficients as well as decision variables are assumed as fuzzy numbers. Ganesan and Veeramani [4] pointed out that for the ranking function $(\Re)$ neither the property $\Re(\tilde{c} \otimes \tilde{x}) = \Re(\tilde{c})\Re(\tilde{x})$ nor $\Re(\tilde{c} \odot \tilde{x}) = \Re(\tilde{c})\Re(\tilde{x})$ is satisfied. To overcome this limitation they have defined a new product $(\odot_{GV})$ for symmetric trapezoidal fuzzy numbers for which the property $\Re(\tilde{c} \odot_{GV} \tilde{x}) = \Re(\tilde{c})\Re(\tilde{x})$ is satisfied and proposed a new method for solving the following type of fuzzy linear programming problems:

$$\text{Maximize/Minimize} \sum_{j=1}^{n} \tilde{c}_j \odot_{GV} \tilde{x}_j$$

subject to

$$\sum_{j=1}^{n} a_{ij}\tilde{x}_j \preceq, \approx, \succeq \tilde{b}_i \ \forall\, i = 1, 2, \ldots, m \tag{5.9}$$

$$\tilde{x}_{ij} \succeq \tilde{0} \ \forall\, j = 1, 2, \ldots, n$$

where $\tilde{x}_j$, $\tilde{c}_j$ and $\tilde{b}_i$ are symmetric trapezoidal fuzzy numbers.

*Example 5.3* [4]

Maximize $((13, 15, 2, 2) \odot_{GV} \tilde{x}_1 \oplus (12, 14, 3, 3) \odot_{GV} \tilde{x}_2 \oplus (15, 17, 2, 2) \odot_{GV} \tilde{x}_3)$
subject to
$$12\tilde{x}_1 + 13\tilde{x}_2 + 12\tilde{x}_3 \preceq (475, 505, 6, 6)$$
$$14\tilde{x}_1 + 13\tilde{x}_3 \preceq (460, 480, 8, 8)$$
$$12\tilde{x}_1 + 15\tilde{x}_2 \preceq (465, 495, 5, 5)$$
$$\tilde{x}_1 \succeq \tilde{0}, \tilde{x}_2 \succeq \tilde{0}, \tilde{x}_3 \succeq \tilde{0}$$

where $\tilde{x}_1$, $\tilde{x}_2$ and $\tilde{x}_3$ are symmetric trapezoidal fuzzy numbers.

(iv) Mahadavi and Nasseri [6] proposed a fuzzy dual simplex method for solving the following type of fuzzy linear programming problems:

$$\text{Minimize} \sum_{j=1}^{n} c_j\tilde{x}_j$$

subject to

$$\sum_{j=1}^{n} a_{ij}\tilde{x}_j \succeq \tilde{b}_i \ \forall\, i = 1, 2, \ldots, m \tag{5.10}$$

$$\tilde{x}_j \succeq \tilde{0}, c_j \geq 0 \ \forall\, j = 1, 2, \ldots, n$$

where $a_{ij}$ is a real number and $\tilde{x}_j$, $\tilde{b}_i$ are trapezoidal fuzzy numbers.

*Example 5.4* [6]

$$\text{Minimize } (6\tilde{x}_1 \oplus 10\tilde{x}_2)$$

subject to

$$2\tilde{x}_1 \oplus 5\tilde{x}_2 \succeq (5, 8, 2, 5)$$

$$3\tilde{x}_1 \oplus 4\tilde{x}_2 \succeq (6, 10, 2, 6)$$

$$\tilde{x}_1 \succeq \tilde{0}, \tilde{x}_2 \succeq \tilde{0}$$

where $\tilde{x}_1$ and $\tilde{x}_2$ are trapezoidal fuzzy numbers.

(v)  The existing method [2] can be used for solving the following type of fully fuzzy linear programming problems:

$$\text{Maximize/Minimize } \sum_{j=1}^{n} \tilde{c}_j \otimes \tilde{x}_j$$

subject to                                                                  (5.11)

$$\sum_{j=1}^{n} \tilde{a}_{ij} \otimes \tilde{x}_j \preceq, \approx, \succeq \tilde{b}_i \ \forall \, i = 1, 2, \ldots, m$$

where $\tilde{x}_j$ is a non-negative *LR* fuzzy number and $\tilde{a}_{ij}$, $\tilde{b}_i$, $\tilde{c}_j$ are non-negative or non-positive *LR* fuzzy numbers.

*Example 5.5* [2]

$$\text{Minimize } ((1, 1, 1)_{LR} \otimes \tilde{x}_1 \oplus (2, 1, 2)_{LR} \otimes \tilde{x}_2)$$

subject to

$$(4, 1, 0)_{LR} \otimes \tilde{x}_1 \oplus (-3, 2, 1)_{LR} \otimes \tilde{x}_2 \succeq (2, 1, 2)_{LR}$$

$$(-3, 1, 2)_{LR} \otimes \tilde{x}_1 \oplus (2, 1, 1)_{LR} \otimes \tilde{x}_2 \succeq (1, 0, 1)_{LR}$$

where $\tilde{x}_1$, $\tilde{x}_2$ are non-negative *LR* fuzzy numbers and $L(x) = R(x) = \max\{0, 1 - x\}$.

## 5.3  Limitations of the Existing Methods

In this section, on the basis of applicability of existing methods, discussed in Sect. 5.2, the limitations of existing methods [2, 4–8] are pointed out.

### 5.3.1  Limitations of the Existing Methods for Solving Fuzzy Linear Programming Problems

In this section, the limitations of the existing methods [4–8] for solving fuzzy linear programming problems are pointed out.

The existing methods [4–8] can be used only to find the fuzzy optimal solution of the following type of fuzzy linear programming problems:

 (i) The existing methods [5, 7] can be used for solving such fuzzy linear programming problems in which the decision variables are represented by crisp numbers and the remaining parameters are represented by triangular or trapezoidal fuzzy numbers.

 (ii) The existing method [8] can be used for solving such fuzzy linear programming problems in which the cost coefficients are represented by triangular or trapezoidal fuzzy numbers and the remaining parameters are represented by crisp numbers.

(iii) The existing method [4] can be used for solving such fuzzy linear programming problems in which the coefficients of the constraints are represented by crisp numbers and the remaining parameters are represented by symmetric triangular or trapezoidal fuzzy numbers.

(iv) The existing method [6] can be used for solving such fuzzy linear programming problems in which the cost coefficients and the coefficients of the constraints are represented by crisp numbers and the remaining parameters are represented by triangular or trapezoidal fuzzy numbers.

However, the existing methods [4–8] cannot be used to find the fuzzy optimal solution of the following type of fuzzy linear programming problems:

(i) The fuzzy linear programming problems in which the decision variables are represented by crisp numbers and the remaining parameters are represented by *LR* fuzzy numbers or *LR* flat fuzzy numbers.

$$\text{Maximize/Minimize} \sum_{j=1}^{n} \tilde{c}_j x_j$$

subject to                                                                                  (5.12)

$$\sum_{j=1}^{n} \tilde{a}_{ij} x_j \preceq, \approx, \succeq \tilde{b}_i \ \forall\, i = 1, 2, \ldots, m$$

$$x_j \geq 0 \ \forall\, j = 1, 2, \ldots, n$$

where $x_j$ is a real number and $\tilde{a}_{ij}, \tilde{b}_i, \tilde{c}_j$ are *LR* flat fuzzy numbers.

*Example 5.6*

$$\text{Maximize} ((2, 3, 3, 1)_{LR} x_1 \oplus (0, 4, 1, 2)_{LR} x_2)$$

subject to

$$(1, 2, 1, 1)_{LR} x_1 \oplus (2, 3, 1, 2)_{LR} x_2 \preceq (5, 6, 2, 2)_{LR}$$
$$(2, 3, 1, 3)_{LR} x_1 \oplus (1, 2, 1, 1)_{LR} x_2 \preceq (4, 6, 2, 1)_{LR}$$
$$x_1, x_2 \geq 0$$

where $L(x) = R(x) = \max\{0, 1 - x^2\}$.

(ii) The fuzzy linear programming problems in which the cost coefficients are represented by *LR* fuzzy numbers or *LR* flat fuzzy numbers and the remaining parameters are represented by crisp numbers.

$$\text{Maximize/Minimize} \sum_{j=1}^{n} \tilde{c}_j x_j$$

subject to

$$\sum_{j=1}^{n} a_{ij} x_j \leq, =, \geq b_i \ \forall i = 1, 2, \ldots, m$$

$$x_j \geq 0 \ \forall j = 1, 2, \ldots, n$$

(5.13)

where $a_{ij}, b_i, x_j$ are real numbers and $\tilde{c}_j$ is an *LR* flat fuzzy number.

*Example 5.7*

$$\text{Maximize} ((3, 8, 5, 4)_{LR} x_1 \oplus (4, 10, 6, 6)_{LR} x_2)$$

subject to

$$2x_1 + 3x_2 \leq 6$$
$$5x_1 + 4x_2 \leq 10$$
$$x_1, x_2 \geq 0$$

where $L(x) = R(x) = \max\{0, 1 - x^4\}$.

(iii) The fuzzy linear programming problems in which the coefficients of the constraints are represented by crisp numbers and the remaining are represented by symmetric *LR* flat fuzzy numbers.

$$\text{Maximize/Minimize} \sum_{j=1}^{n} \tilde{c}_j \otimes \tilde{x}_j$$

subject to

(5.14)

$$\sum_{j=1}^{n} (a_{ij}\tilde{x}_j) \preceq, \approx, \succeq \tilde{b}_i \ \forall i = 1, 2, \ldots, m$$

$$\tilde{x}_j \succeq \tilde{0} \ \forall j = 1, 2, \ldots, n$$

where $\tilde{x}_j$, $\tilde{c}_j$ and $\tilde{b}_i$ are symmetric $LR$ flat fuzzy numbers.

*Example 5.8*

$$\text{Maximize} ((2, 5, 3, 3)_{LR} \otimes \tilde{x}_1 \oplus (3, 4, 4, 4)_{LR} \otimes \tilde{x}_2)$$

subject to

$$2\tilde{x}_1 + 3\tilde{x}_2 \preceq (10, 12, 2, 2)_{LR}$$

$$3\tilde{x}_1 + 4\tilde{x}_2 \preceq (20, 30, 5, 5)_{LR}$$

where $\tilde{x}_1$, $\tilde{x}_2$ are non-negative symmetric trapezoidal fuzzy numbers and $L(x) = R(x) = \max\{0, 1 - x^4\}$.

(iv) The fuzzy linear programming problems in which the cost coefficients and the coefficients of the constraints are represented by crisp numbers and the remaining parameters are represented by $LR$ fuzzy numbers or $LR$ flat fuzzy numbers.

$$\text{Minimize} \sum_{j=1}^{n} c_j \tilde{x}_j$$

subject to

(5.15)

$$\sum_{j=1}^{n} a_{ij}\tilde{x}_j \succeq \tilde{b}_i \ \forall i = 1, 2, \ldots, m$$

$$\tilde{x}_j \succeq \tilde{0}, c_j \geq 0 \ \forall j = 1, 2, \ldots, n$$

where $a_{ij}$ is a real number and $\tilde{x}_j$, $\tilde{b}_i$ are $LR$ flat fuzzy numbers.

*Example 5.9*

$$\text{Minimize} (6\tilde{x}_1 \oplus 10\tilde{x}_2)$$

subject to

$$2\tilde{x}_1 \oplus 5\tilde{x}_2 \succeq (5, 8, 6, 5)_{LR}$$

$$3\tilde{x}_1 \oplus 4\tilde{x}_2 \succeq (6, 10, 2, 6)_{LR}$$

$$\tilde{x}_1 \succeq \tilde{0}, \tilde{x}_2 \succeq \tilde{0}$$

where $\tilde{x}_1$, $\tilde{x}_2$ are $LR$ flat fuzzy numbers and $L(x) = R(x) = \max\{0, 1 - x^2\}$.

### 5.3.2   Limitations of the Existing Method for Solving Fully Fuzzy Linear Programming Problems

In this section, the limitations of the existing method [2] for solving fully fuzzy linear programming problems with inequality constraints are pointed out:

The existing method [2] can be used for solving such fully fuzzy linear programming problems with inequality constraints in which all the coefficients are represented by either non-negative $LR$ fuzzy numbers or non-positive $LR$ fuzzy numbers and all the decision variables are represented by non-negative $LR$ fuzzy numbers.

However, the existing method [2] cannot be used to find the fuzzy optimal solution of such fully fuzzy linear programming problems in which some or all the parameters are represented by such $LR$ fuzzy numbers or $LR$ flat fuzzy numbers which are neither non-negative nor non-positive.

$$\text{Maximize/Minimize} \sum_{j=1}^{n} \tilde{c}_j \otimes \tilde{x}_j$$

subject to                                                                    (5.16)

$$\sum_{j=1}^{n} \tilde{a}_{ij} \otimes \tilde{x}_j \preceq, =, \succeq \tilde{b}_i \ \forall i = 1, 2, \ldots, m$$

where $\tilde{a}_{ij}, \tilde{x}_j, \tilde{b}_i$ and $\tilde{c}_j$ are $LR$ flat fuzzy numbers.

*Example 5.10*

Maximize $((4, 4, 0, 0)_{LR} \otimes \tilde{x}_1 \oplus (1, 1, 1, 1)_{LR} \otimes \tilde{x}_2)$

subject to

$(2, 5, 5, 2)_{LR} \otimes \tilde{x}_1 \oplus (-1, 5, 1, 2)_{LR} \otimes \tilde{x}_2 \preceq (-17, 45, 25, 46)_{LR}$

$(1, 2, 1, 1)_{LR} \otimes \tilde{x}_1 \oplus (3, 5, 2, 2)_{LR} \otimes \tilde{x}_2 = (7, 24, 6, 24)_{LR}$

where $\tilde{x}_1, \tilde{x}_2$ are $LR$ flat fuzzy numbers and $L(x) = \max\{0, 1-x\}, R(x) = \max\{0, 1 - x^2\}$.

$$\text{Maximize/Minimize} \sum_{j=1}^{n} \tilde{c}_j \odot \tilde{x}_j$$

subject to                                                                    (5.17)

$$\sum_{j=1}^{n} \tilde{a}_{ij} \odot \tilde{x}_j \preceq, =, \succeq \tilde{b}_i \ \forall i = 1, 2, \ldots, m$$

where $\tilde{a}_{ij}, \tilde{x}_j, \tilde{b}_i$ and $\tilde{c}_j$ are $LR$ flat fuzzy numbers.

*Example 5.11*

Maximize $((1, 1, 1, 1)_{LR} \odot \tilde{x}_1 \oplus (4, 4, 0, 0)_{LR} \odot \tilde{x}_2)$

subject to

$(2, 3, 1, 1)_{LR} \odot \tilde{x}_1 \oplus (-3, -2, 1, 1)_{LR} \odot \tilde{x}_2 \succeq (-27, -4, 21, 32)_{LR}$

$(1, 2, 1, 1)_{LR} \odot \tilde{x}_1 \oplus (3, 5, 2, 2)_{LR} \odot \tilde{x}_2 = (11, 39, \dfrac{193}{5}, \dfrac{168}{5})_{LR}$

where $\tilde{x}_1, \tilde{x}_2$ are *LR* flat fuzzy numbers and $L(x) = R(x) = \max\{0, 1 - x\}$.


## 5.4 Kumar and Kaur's Methods for Solving Fully Fuzzy Linear Programming Problems with Inequality Constraints Having *LR* Flat Fuzzy Numbers

On the basis of the limitations of the existing methods [2, 4–8], discussed in Sect. 5.3, it can be concluded that there was no method in literature to find the fuzzy optimal solution of such fuzzy linear programming problems and fully fuzzy linear programming problems in which some of all the parameters are represented by such *LR* fuzzy numbers or *LR* flat fuzzy numbers which are neither non-negative nor non-positive. Therefore, to overcome the limitations of the existing methods two new methods, proposed by Kumar and Kaur [1] are presented in this section to find the fuzzy optimal solution of fully fuzzy linear programming problems with inequality constraints.


### 5.4.1 Kumar and Kaur's Method

In this section, a method proposed by Kumar and Kaur [1] to find the fuzzy optimal solution of fully fuzzy linear programming problems of type (5.16) with inequality constraints, is presented. The same method can also be used to find the fuzzy optimal solution of the proposed fully fuzzy linear programming problems (5.17) by replacing $\otimes$ by $\odot$ and also other fuzzy linear programming problems (5.7)–(5.15) by replacing the fuzzy parameters by crisp parameters.

The steps of the method for solving fully fuzzy linear programming problems (5.16) are as follows:

**Step 1** Dividing all the constraints into three categories i.e., $\sum\limits_{j=1}^{n} \tilde{a}_{pj} \otimes \tilde{x}_j \preceq \tilde{b}_p \; \forall p \in N_1$, $\sum\limits_{j=1}^{n} \tilde{a}_{qj} \otimes \tilde{x}_j \succeq \tilde{b}_q \; \forall q \in N_2$ and $\sum\limits_{j=1}^{n} \tilde{a}_{rj} \otimes \tilde{x}_j = \tilde{b}_r \; \forall r \in N_3$ the fully fuzzy linear programming problem (5.16) can be written as:

$$\text{Maximize/Minimize} \sum_{j=1}^{n} \tilde{c}_j \otimes \tilde{x}_j$$

subject to

$$\sum_{j=1}^{n} \tilde{a}_{pj} \otimes \tilde{x}_j \preceq \tilde{b}_p \qquad \forall \, p \in N_1$$

$$\sum_{j=1}^{n} \tilde{a}_{qj} \otimes \tilde{x}_j \succeq \tilde{b}_q \qquad \forall \, q \in N_2 \qquad\qquad (5.18)$$

$$\sum_{j=1}^{n} \tilde{a}_{rj} \otimes \tilde{x}_j = \tilde{b}_r \qquad \forall \, r \in N_3$$

where $\tilde{x}_j$ is an *LR* flat fuzzy number and $N_1 = \{i \, : \, 1 \leq i \leq m$ and $\sum_{j=1}^{n} \tilde{a}_{ij} \otimes \tilde{x}_j \preceq \tilde{b}_i\}$,

$N_2 = \{i \, : \, 1 \leq i \leq m$ and $\sum_{j=1}^{n} \tilde{a}_{ij} \otimes \tilde{x}_j \succeq \tilde{b}_i\}$, $N_3 = \{i \, : \, 1 \leq i \leq m$ and $\sum_{j=1}^{n} \tilde{a}_{ij}$

$\otimes \tilde{x}_j = \tilde{b}_i\}$.

**Step 2** Convert the inequality constraints $\sum_{j=1}^{n} \tilde{a}_{pj} \otimes \tilde{x}_j \preceq \tilde{b}_p \; \forall \, p \in N_1$ into equality constraints by introducing fuzzy variable $\tilde{S}_p$ to the left side and $\tilde{S}'_p$ to the right side of the constraint i.e.,

$$\sum_{j=1}^{n} \tilde{a}_{pj} \otimes \tilde{x}_j \oplus \tilde{S}_p = \tilde{b}_p \oplus \tilde{S}'_p \; \forall \, p \in N_1$$

where $\Re(\tilde{S}_p) - \Re(\tilde{S}'_p) \geq 0$.

Convert the inequality constraints $\sum_{j=1}^{n} \tilde{a}_{qj} \otimes \tilde{x}_j \succeq \tilde{b}_q \; \forall \, q \in N_2$ into equality constraints by introducing fuzzy variable $\tilde{S}_q$ to the left side and $\tilde{S}'_q$ to the right side of the constraint i.e.,

$$\sum_{j=1}^{n} \tilde{a}_{qj} \otimes \tilde{x}_j \oplus \tilde{S}_q = \tilde{b}_q \oplus \tilde{S}'_q \; \forall \, q \in N_2$$

where $\Re(\tilde{S}_q) - \Re(\tilde{S}'_q) \leq 0$.

**Step 3** The fully fuzzy linear programming problem, obtained from Step 2, can be written as:

$$\text{Maximize/Minimize} \sum_{j=1}^{n} \tilde{c}_j \otimes \tilde{x}_j$$

subject to

$$\sum_{j=1}^{n} \tilde{a}_{pj} \otimes \tilde{x}_j \oplus \tilde{S}_p = \tilde{b}_p \oplus \tilde{S}'_p \quad \forall\, p \in N_1$$

$$\sum_{j=1}^{n} \tilde{a}_{qj} \otimes \tilde{x}_j \oplus \tilde{S}_q = \tilde{b}_q \oplus \tilde{S}'_q \quad \forall\, q \in N_2 \tag{5.19}$$

$$\sum_{j=1}^{n} \tilde{a}_{rj} \otimes \tilde{x}_j = \tilde{b}_r \quad \forall\, r \in N_3$$

$$\Re(\tilde{S}_p) - \Re(\tilde{S}'_p) \geq 0 \quad \forall\, p \in N_1$$

$$\Re(\tilde{S}_q) - \Re(\tilde{S}'_q) \leq 0 \quad \forall\, q \in N_2$$

where $\tilde{x}_j$, $\tilde{S}_p$, $\tilde{S}'_p$, $\tilde{S}_q$ and $\tilde{S}'_q$ are *LR* flat fuzzy number.

**Step 4** Assuming $\tilde{a}_{ij} = (a_{ij}, b_{ij}, \alpha_{ij}, \beta_{ij})_{LR}$, $\tilde{x}_j = (x_j, y_j, \alpha''_j, \beta''_j)_{LR}$, $\tilde{b}_i = (b_i, g_i, \gamma_i, \delta_i)_{LR}$, $\tilde{c}_j = (p_j, q_j, \alpha'_j, \beta'_j)_{LR}$, $\tilde{S}_i = (s_i, t_i, \eta_i, \rho_i)_{LR}$ and $\tilde{S}'_i = (s'_i, t'_i, \eta'_i, \rho'_i)_{LR}$ the fully fuzzy linear programming problem (5.19) can be written as:

$$\text{Maximize/Minimize} \sum_{j=1}^{n} (p_j, q_j, \alpha'_j, \beta'_j)_{LR} \otimes (x_j, y_j, \alpha''_j, \beta''_j)_{LR}$$

subject to

$$\sum_{j=1}^{n} (a_{pj}, b_{pj}, \alpha_{pj}, \beta_{pj})_{LR} \otimes (x_j, y_j, \alpha''_j, \beta''_j)_{LR} \oplus (s_p, t_p, \eta_p, \rho_p)_{LR} = (b_p, g_p, \gamma_p,$$
$$\delta_p)_{LR} \oplus (s'_p, t'_p, \eta'_p, \rho'_p)_{LR} \qquad\qquad \forall\, p \in N_1$$

$$\sum_{j=1}^{n} (a_{qj}, b_{qj}, \alpha_{qj}, \beta_{qj})_{LR} \otimes (x_j, y_j, \alpha''_j, \beta''_j)_{LR} \oplus (s_q, t_q, \eta_q, \rho_q)_{LR} = (b_q, g_q, \gamma_q, \tag{5.20}$$
$$\delta_q)_{LR} \oplus (s'_q, t'_q, \eta'_q, \rho'_q)_{LR} \qquad\qquad \forall\, q \in N_2$$

$$\sum_{j=1}^{n} (a_{rj}, b_{rj}, \alpha_{rj}, \beta_{rj})_{LR} \otimes (x_j, y_j, \alpha''_j, \beta''_j)_{LR} = (b_r, g_r, \gamma_r, \delta_r)_{LR} \quad \forall\, r \in N_3$$

$$\Re(s_p, t_p, \eta_p, \rho_p)_{LR} - \Re(s'_p, t'_p, \eta'_p, \rho'_p)_{LR} \geq 0 \quad \forall\, p \in N_1$$

$$\Re(s_q, t_q, \eta_q, \rho_q)_{LR} - \Re(s'_q, t'_q, \eta'_q, \rho'_q)_{LR} \leq 0 \quad \forall\, q \in N_2$$

where $(x_j, y_j, \alpha''_j, \beta''_j)_{LR}$, $(s_p, t_p, \eta_p, \rho_p)_{LR}$, $(s'_p, t'_p, \eta'_p, \rho'_p)_{LR}$, $(s_q, t_q, \eta_q, \rho_q)_{LR}$ and $(s'_q, t'_q, \eta'_q, \rho'_q)_{LR}$ are *LR* flat fuzzy number.

**Step 5** Using the product, presented in Sect. 4.2.1 and assuming $(a_{ij}, b_{ij}, \alpha_{ij}, \beta_{ij})_{LR} \otimes (x_j, y_j, \alpha''_j, \beta''_j)_{LR} = (m_{ij}, n_{ij}, \gamma'_{ij}, \delta'_{ij})_{LR}$ the fully fuzzy linear programming problem (5.20) can be written as:

$$\text{Maximize/Minimize} \sum_{j=1}^{n} (p_j, q_j, \alpha'_j, \beta'_j)_{LR} \otimes (x_j, y_j, \alpha''_j, \beta''_j)_{LR}$$

subject to

$$\sum_{j=1}^{n} (m_{pj}, n_{pj}, \gamma'_{pj}, \delta'_{pj})_{LR} \oplus (s_p, t_p, \eta_p, \rho_p)_{LR} = (b_p, g_p, \gamma_p, \delta_p)_{LR} \oplus (s'_p, t'_p,$$

$$\eta'_p, \rho'_p)_{LR} \qquad\qquad\qquad\qquad \forall\, p \in N_1$$

$$\sum_{j=1}^{n} (m_{qj}, n_{qj}, \gamma'_{qj}, \delta'_{qj})_{LR} \oplus (s_q, t_q, \eta_q, \rho_q)_{LR} = (b_q, g_q, \gamma_q, \delta_q)_{LR} \oplus (s'_q, t'_q,$$

$$\eta'_q, \rho'_q)_{LR} \qquad\qquad\qquad\qquad \forall\, q \in N_2$$

$$\sum_{j=1}^{n} (m_{rj}, n_{rj}, \gamma'_{rj}, \delta'_{rj})_{LR} = (b_r, g_r, \gamma_r, \delta_r)_{LR} \qquad\qquad \forall\, r \in N_3$$

$$\Re(s_p, t_p, \eta_p, \rho_p)_{LR} - \Re(s'_p, t'_p, \eta'_p, \rho'_p)_{LR} \geq 0 \qquad\qquad \forall\, p \in N_1$$

$$\Re(s_q, t_q, \eta_q, \rho_q)_{LR} - \Re(s'_q, t'_q, \eta'_q, \rho'_q)_{LR} \leq 0 \qquad\qquad \forall\, q \in N_2$$

(5.21)

where $(x_j, y_j, \alpha''_j, \beta''_j)_{LR}$, $(s_p, t_p, \eta_p, \rho_p)_{LR}$, $(s'_p, t'_p, \eta'_p, \rho'_p)_{LR}$, $(s_q, t_q, \eta_q, \rho_q)_{LR}$ and $(s'_q, t'_q, \eta'_q, \rho'_q)_{LR}$ are $LR$ flat fuzzy number.

**Step 6** Using arithmetic operations, defined in Sect. 4.1.2 and Definition 4.5, the fully fuzzy linear programming problem (5.21) can be written as:

$$\text{Maximize/Minimize} \sum_{j=1}^{n} (p_j, q_j, \alpha'_j, \beta'_j)_{LR} \otimes (x_j, y_j, \alpha''_j, \beta''_j)_{LR}$$

subject to

$$\sum_{j=1}^{n} m_{pj} + s_p = b_p + s'_p \ \forall\, p \in N_1, \qquad\qquad \sum_{j=1}^{n} n_{pj} + t_p = g_p + t'_p \ \forall\, p \in N_1$$

$$\sum_{j=1}^{n} \gamma'_{pj} + \eta_p = \gamma_p + \eta'_p \ \forall\, p \in N_1, \qquad\qquad \sum_{j=1}^{n} \delta'_{pj} + \rho_p = \delta_p + \rho'_p \ \forall\, p \in N_1$$

(5.22)

$$\sum_{j=1}^{n} m_{qj} + s_q = b_q + s'_q \ \forall\, q \in N_2, \qquad\qquad \sum_{j=1}^{n} n_{qj} + t_q = g_q + t'_q \ \forall\, q \in N_2$$

$$\sum_{j=1}^{n} \gamma'_{qj} + \eta_q = \gamma_q + \eta'_q \ \forall q \in N_2, \qquad \sum_{j=1}^{n} \delta'_{qj} + \rho_q = \delta_q + \rho'_q \ \forall q \in N_2$$

$$\sum_{j=1}^{n} m_{rj} = b_r \qquad \forall r \in N_3, \qquad \sum_{j=1}^{n} n_{rj} = g_r \qquad \forall r \in N_3$$

$$\sum_{j=1}^{n} \gamma'_{rj} = \gamma_r \qquad \forall r \in N_3, \qquad \sum_{j=1}^{n} \delta'_{rj} = \delta_r \qquad \forall r \in N_3$$

$$\Re(s_p, t_p, \eta_p, \rho_p)_{LR} - \Re(s'_p, t'_p, \eta'_p, \rho'_p)_{LR} \geq 0 \ \forall p \in N_1$$

$$\Re(s_q, t_q, \eta_q, \rho_q)_{LR} - \Re(s'_q, t'_q, \eta'_q, \rho'_q)_{LR} \leq 0 \ \forall q \in N_2$$

$$x_j \leq y_j, \alpha''_j \geq 0, \beta''_j \geq 0, s_i \leq t_i, \eta_i \geq 0, \rho_i \geq 0, s'_i \leq t'_i, \eta'_i \geq 0, \rho'_i \geq 0$$

$$\forall i = 1, 2, \ldots, m; \ j = 1, 2, \ldots, n$$

**Step 7** As discussed in Step 4 of Sect. 4.4 the fuzzy optimal solution of (5.16) can be obtained by solving the problem (5.23):

$$\text{Maximize/Minimize } \Re(\sum_{j=1}^{n} (p_j, q_j, \alpha'_j, \beta'_j)_{LR} \otimes (x_j, y_j, \alpha''_j, \beta''_j)_{LR})$$

subject to

$$\sum_{j=1}^{n} m_{pj} + s_p = b_p + s'_p \ \forall p \in N_1, \qquad \sum_{j=1}^{n} n_{pj} + t_p = g_p + t'_p \ \forall p \in N_1$$

$$\sum_{j=1}^{n} \gamma'_{pj} + \eta_p = \gamma_p + \eta'_p \ \forall p \in N_1, \qquad \sum_{j=1}^{n} \delta'_{pj} + \rho_p = \delta_p + \rho'_p \ \forall p \in N_1$$

$$\sum_{j=1}^{n} m_{qj} + s_q = b_q + s'_q \ \forall q \in N_2, \qquad \sum_{j=1}^{n} n_{qj} + t_q = g_q + t'_q \ \forall q \in N_2$$

$$\sum_{j=1}^{n} \gamma'_{qj} + \eta_q = \gamma_q + \eta'_q \ \forall q \in N_2, \qquad \sum_{j=1}^{n} \delta'_{qj} + \rho_q = \delta_q + \rho'_q \ \forall q \in N_2 \tag{5.23}$$

$$\sum_{j=1}^{n} m_{rj} = b_r \qquad \forall r \in N_3, \qquad \sum_{j=1}^{n} n_{rj} = g_r \qquad \forall r \in N_3$$

$$\sum_{j=1}^{n} \gamma'_{rj} = \gamma_r \qquad \forall r \in N_3, \qquad \sum_{j=1}^{n} \delta'_{rj} = \delta_r \qquad \forall r \in N_3$$

$$\Re(s_p, t_p, \eta_p, \rho_p)_{LR} - \Re(s'_p, t'_p, \eta'_p, \rho'_p)_{LR} \geq 0 \ \forall p \in N_1$$

$$\Re(s_q, t_q, \eta_q, \rho_q)_{LR} - \Re(s'_q, t'_q, \eta'_q, \rho'_q)_{LR} \leq 0 \ \forall q \in N_2$$

$$x_j \leq y_j, \alpha''_j \geq 0, \beta''_j \geq 0, s_i \leq t_i, \eta_i \geq 0, \rho_i \geq 0, s'_i \leq t'_i, \eta'_i \geq 0, \rho'_i \geq 0$$

$$\forall i = 1, 2, \ldots, m; \ j = 1, 2, \ldots, n$$

**Step 8** Assuming $(p_j, q_j, \alpha'_j, \beta'_j)_{LR} \otimes (x_j, y_j, \alpha''_j, \beta''_j)_{LR} = (p'_j, q'_j, \alpha'''_j, \beta'''_j)_{LR}$ the problem (5.23) can be written as:

$$\text{Maximize/Minimize } \Re(\sum_{j=1}^{n} (p'_j, q'_j, \alpha'''_j, \beta'''_j)_{LR})$$

subject to

$$\sum_{j=1}^{n} m_{pj} + s_p = b_p + s'_p \ \forall p \in N_1, \qquad \sum_{j=1}^{n} n_{pj} + t_p = g_p + t'_p \ \forall p \in N_1$$

$$\sum_{j=1}^{n} \gamma'_{pj} + \eta_p = \gamma_p + \eta'_p \ \forall p \in N_1, \qquad \sum_{j=1}^{n} \delta'_{pj} + \rho_p = \delta_p + \rho'_p \ \forall p \in N_1$$

$$\sum_{j=1}^{n} m_{qj} + s_q = b_q + s'_q \ \forall q \in N_2, \qquad \sum_{j=1}^{n} n_{qj} + t_q = g_q + t'_q \ \forall q \in N_2$$

$$\sum_{j=1}^{n} \gamma'_{qj} + \eta_q = \gamma_q + \eta'_q \ \forall q \in N_2, \qquad \sum_{j=1}^{n} \delta'_{qj} + \rho_q = \delta_q + \rho'_q \ \forall q \in N_2 \qquad (5.24)$$

$$\sum_{j=1}^{n} m_{rj} = b_r \qquad \forall r \in N_3, \qquad \sum_{j=1}^{n} n_{rj} = g_r \qquad \forall r \in N_3$$

$$\sum_{j=1}^{n} \gamma'_{rj} = \gamma_r \qquad \forall r \in N_3, \qquad \sum_{j=1}^{n} \delta'_{rj} = \delta_r \qquad \forall r \in N_3$$

$$\Re(s_p, t_p, \eta_p, \rho_p)_{LR} - \Re(s'_p, t'_p, \eta'_p, \rho'_p)_{LR} \geq 0 \ \forall p \in N_1$$

$$\Re(s_q, t_q, \eta_q, \rho_q)_{LR} - \Re(s'_q, t'_q, \eta'_q, \rho'_q)_{LR} \leq 0 \ \forall q \in N_2$$

$$x_j \leq y_j, \alpha''_j \geq 0, \beta''_j \geq 0, s_i \leq t_i, \eta_i \geq 0, \rho_i \geq 0, s'_i \leq t'_i, \eta'_i \geq 0, \rho'_i \geq 0$$

$$\forall i = 1, 2, \ldots, m; \ j = 1, 2, \ldots, n$$

**Step 9** Using the linearity property $\Re(\sum_{j=1}^{n} \tilde{A}_i) = \sum_{j=1}^{n} \Re(\tilde{A}_i)$, where $\tilde{A}_i$ is a fuzzy number, the problem (5.24) can be converted into (5.25):

$$\text{Maximize/Minimize } \sum_{j=1}^{n} \Re(p'_j, q'_j, \alpha'''_j, \beta'''_j)_{LR}$$

subject to

$$\sum_{j=1}^{n} m_{pj} + s_p = b_p + s'_p \ \forall p \in N_1, \qquad \sum_{j=1}^{n} n_{pj} + t_p = g_p + t'_p \ \forall p \in N_1$$

$$\sum_{j=1}^{n} \gamma'_{pj} + \eta_p = \gamma_p + \eta'_p \ \forall \, p \in N_1, \qquad \sum_{j=1}^{n} \delta'_{pj} + \rho_p = \delta_p + \rho'_p \ \forall \, p \in N_1$$

$$(5.25)$$

$$\sum_{j=1}^{n} m_{qj} + s_q = b_q + s'_q \ \forall \, q \in N_2, \qquad \sum_{j=1}^{n} n_{qj} + t_q = g_q + t'_q \ \forall \, q \in N_2$$

$$\sum_{j=1}^{n} \gamma'_{qj} + \eta_q = \gamma_q + \eta'_q \ \forall \, q \in N_2, \qquad \sum_{j=1}^{n} \delta'_{qj} + \rho_q = \delta_q + \rho'_q \ \forall \, q \in N_2$$

$$\sum_{j=1}^{n} m_{rj} = b_r \qquad \forall \, r \in N_3, \qquad \sum_{j=1}^{n} n_{rj} = g_r \qquad \forall \, r \in N_3$$

$$\sum_{j=1}^{n} \gamma'_{rj} = \gamma_r \qquad \forall \, r \in N_3, \qquad \sum_{j=1}^{n} \delta'_{rj} = \delta_r \qquad \forall \, r \in N_3$$

$$\Re(s_p, t_p, \eta_p, \rho_p)_{LR} - \Re(s'_p, t'_p, \eta'_p, \rho'_p)_{LR} \geq 0 \ \forall \, p \in N_1$$

$$\Re(s_q, t_q, \eta_q, \rho_q)_{LR} - \Re(s'_q, t'_q, \eta'_q, \rho'_q)_{LR} \leq 0 \ \forall \, q \in N_2$$

$$x_j \leq y_j, \alpha''_j \geq 0, \beta''_j \geq 0, s_i \leq t_i, \eta_i \geq 0, \rho_i \geq 0, s'_i \leq t'_i, \eta'_i \geq 0, \rho'_i \geq 0$$

$$\forall \, i = 1, 2, \ldots, m; \ j = 1, 2, \ldots, n$$

**Step 10** Using $\Re(m, n, \alpha, \beta) = \frac{1}{2}(\int_0^1 (m - \alpha L^{-1}(\lambda)) \, d\lambda + \int_0^1 (n + \beta R^{-1}(\lambda)) \, d\lambda)$ problem (5.25) can be converted into (5.26):

$$\text{Maximize/Minimize} \sum_{j=1}^{n} (\frac{1}{2}(\int_0^1 (p'_j - \alpha'''_j L^{-1}(\lambda)) d\lambda + \int_0^1 (q'_j + \beta'''_j R^{-1}(\lambda)) d\lambda))$$

subject to

$$\sum_{j=1}^{n} m_{pj} + s_p = b_p + s'_p \ \forall \, p \in N_1, \qquad \sum_{j=1}^{n} n_{pj} + t_p = g_p + t'_p \ \forall \, p \in N_1$$

$$\sum_{j=1}^{n} \gamma'_{pj} + \eta_p = \gamma_p + \eta'_p \ \forall \, p \in N_1, \qquad \sum_{j=1}^{n} \delta'_{pj} + \rho_p = \delta_p + \rho'_p \ \forall \, p \in N_1$$

$$(5.26)$$

$$\sum_{j=1}^{n} m_{qj} + s_q = b_q + s'_q \ \forall \, q \in N_2, \qquad \sum_{j=1}^{n} n_{qj} + t_q = g_q + t'_q \ \forall \, q \in N_2$$

$$\sum_{j=1}^{n} \gamma'_{qj} + \eta_q = \gamma_q + \eta'_q \ \forall \, q \in N_2, \qquad \sum_{j=1}^{n} \delta'_{qj} + \rho_q = \delta_q + \rho'_q \ \forall \, q \in N_2$$

$$\sum_{j=1}^{n} m_{rj} = b_r \qquad \forall\, r \in N_3, \qquad \sum_{j=1}^{n} n_{rj} = g_r \qquad \forall\, r \in N_3$$

$$\sum_{j=1}^{n} \gamma'_{rj} = \gamma_r \qquad \forall\, r \in N_3, \qquad \sum_{j=1}^{n} \delta'_{rj} = \delta_r \qquad \forall\, r \in N_3$$

$$\frac{1}{2}\Big(\int_0^1 (s_p - \eta_p L^{-1}(\lambda))d\lambda + \int_0^1 (t_p + \rho_p R^{-1}(\lambda))d\lambda\Big) - \frac{1}{2}\Big(\int_0^1 (s'_p - \eta'_p L^{-1}(\lambda))d\lambda +$$

$$\int_0^1 (t'_p + \rho'_p R^{-1}(\lambda))d\lambda\Big) \geq 0 \;\; \forall\, p \in N_1$$

$$\frac{1}{2}\Big(\int_0^1 (s_q - \eta_q L^{-1}(\lambda))d\lambda + \int_0^1 (t_q + \rho_q R^{-1}(\lambda))d\lambda\Big) - \frac{1}{2}\Big(\int_0^1 (s'_q - \eta'_q L^{-1}(\lambda))d\lambda +$$

$$\int_0^1 (t'_q + \rho'_q R^{-1}(\lambda))d\lambda\Big) \leq 0 \;\; \forall\, p \in N_2$$

$$x_j \leq y_j, \alpha''_j \geq 0, \beta''_j \geq 0, s_i \leq t_i, \eta_i \geq 0, \rho_i \geq 0, s'_i \leq t'_i, \eta'_i \geq 0, \rho'_i \geq 0 \;\forall\, i =$$
$$1, 2, \ldots, m; \; j = 1, 2, \ldots, n$$

**Step 11** Solve the crisp non-linear programming problem (5.26) by using an appropriate existing method [9] to find the optimal solution $\{x_j^*, y_j^*, \alpha_j''^*, \beta_j''^*\}$.

**Step 12** Find the fuzzy optimal solution $\{\tilde{x}_j^*\}$ of the fully fuzzy linear programming problem (5.16) by putting the values of $x_j^*, y_j^*, \alpha_j''^*$ and $\beta_j''^*$ in $\tilde{x}_j^* = (x_j^*, y_j^*, \alpha_j''^*, \beta_j''^*)$.

**Step 13** Find the fuzzy optimal value by putting the values of $\tilde{x}_j^*$, obtained from Step 12, in $\sum_{j=1}^{n} \tilde{c}_j \otimes \tilde{x}_j^*$.

### 5.4.2 Alternative Method

In this section, an alternative method [1], to find the fuzzy optimal solution of fully fuzzy linear programming problems of type (5.16) with inequality constraints, is presented. The same method can be used to find the fuzzy optimal solution of the fully fuzzy linear programming problems and fuzzy linear programming problems which can be solved by using the method, presented in Sect. 5.4.1.

The steps of the method for solving fully fuzzy linear programming problem (5.16) are as follows:

**Step 1** Assuming $\tilde{a}_{ij} = (a_{ij}, b_{ij}, \alpha_{ij}, \beta_{ij})_{LR}$, $\tilde{x}_j = (x_j, y_j, \alpha''_j, \beta''_j)_{LR}$, $\tilde{b}_i = (b_i, g_i, \gamma_i, \delta_i)_{LR}$ and $\tilde{c}_j = (p_j, q_j, \alpha'_j, \beta'_j)_{LR}$ the fully fuzzy linear programming problem (5.18) can be written as:

$$\text{Maximize/Minimize} \sum_{j=1}^{n} (p_j, q_j, \alpha'_j, \beta'_j)_{LR} \otimes (x_j, y_j, \alpha''_j, \beta''_j)_{LR}$$

subject to

$$\sum_{j=1}^{n} (a_{pj}, b_{pj}, \alpha_{pj}, \beta_{pj})_{LR} \otimes (x_j, y_j, \alpha''_j, \beta''_j)_{LR} \preceq (b_p, g_p, \gamma_p, \delta_p)_{LR} \quad \forall \, p \in N_1$$

$$\sum_{j=1}^{n} (a_{qj}, b_{qj}, \alpha_{qj}, \beta_{qj})_{LR} \otimes (x_j, y_j, \alpha''_j, \beta''_j)_{LR} \succeq (b_q, g_q, \gamma_q, \delta_q)_{LR} \quad \forall \, q \in N_2 \tag{5.27}$$

$$\sum_{j=1}^{n} (a_{rj}, b_{rj}, \alpha_{rj}, \beta_{rj})_{LR} \otimes (x_j, y_j, \alpha''_j, \beta''_j)_{LR} = (b_r, g_r, \gamma_r, \delta_r)_{LR} \quad \forall \, r \in N_3$$

where $(x_j, y_j, \alpha''_j, \beta''_j)_{LR}$ is an *LR* flat fuzzy number.

**Step 2** Using the product, presented in Sect. 4.1.2 and assuming $(a_{ij}, b_{ij}, \alpha_{ij}, \beta_{ij})_{LR} \otimes (x_j, y_j, \alpha''_j, \beta''_j)_{LR} = (m_{ij}, n_{ij}, \gamma'_{ij}, \delta'_{ij})_{LR}$ the fully fuzzy linear programming problem (5.27) can be written as:

$$\text{Maximize/Minimize} \sum_{j=1}^{n} (p_j, q_j, \alpha'_j, \beta'_j)_{LR} \otimes (x_j, y_j, \alpha''_j, \beta''_j)_{LR}$$

subject to

$$\sum_{j=1}^{n} (m_{pj}, n_{pj}, \gamma'_{pj}, \delta'_{pj})_{LR} \preceq (b_p, g_p, \gamma_p, \delta_p)_{LR} \quad \forall \, p \in N_1$$

$$\sum_{j=1}^{n} (m_{qj}, n_{qj}, \gamma'_{qj}, \delta'_{qj})_{LR} \succeq (b_q, g_q, \gamma_q, \delta_q)_{LR} \quad \forall \, q \in N_2 \tag{5.28}$$

$$\sum_{j=1}^{n} (m_{rj}, n_{rj}, \gamma'_{rj}, \delta'_{rj})_{LR} = (b_r, g_r, \gamma_r, \delta_r)_{LR} \quad \forall \, r \in N_3$$

where $(x_j, y_j, \alpha''_j, \beta''_j)_{LR}$ is an *LR* flat fuzzy number.

**Step 3** Using arithmetic operations, defined in Sect. 4.1.2 and Definition 4.5, the fully fuzzy linear programming problem (5.28) can be written as:

$$\text{Maximize/Minimize} \sum_{j=1}^{n} (p_j, q_j, \alpha'_j, \beta'_j)_{LR} \otimes (x_j, y_j, \alpha''_j, \beta''_j)_{LR}$$

subject to

$$\sum_{j=1}^{n} (m_{pj}, n_{pj}, \gamma'_{pj}, \delta'_{pj})_{LR} \preceq (b_p, g_p, \gamma_p, \delta_p)_{LR} \quad \forall p \in N_1$$

$$\sum_{j=1}^{n} (m_{qj}, n_{qj}, \gamma'_{qj}, \delta'_{qj})_{LR} \succeq (b_q, g_q, \gamma_q, \delta_q)_{LR} \quad \forall q \in N_2 \qquad (5.29)$$

$$\sum_{j=1}^{n} m_{rj} = b_r \quad \forall r \in N_3, \qquad \sum_{j=1}^{n} n_{rj} = g_r \qquad \forall r \in N_3$$

$$\sum_{j=1}^{n} \gamma'_{rj} = \gamma_r \quad \forall r \in N_3, \qquad \sum_{j=1}^{n} \delta'_{rj} = \delta_r \qquad \forall r \in N_3$$

$$x_j \leq y_j, \alpha''_j \geq 0, \beta''_j \geq 0 \; \forall j = 1, 2, \ldots, n$$

**Step 4** As discussed in Step 4 of the Sect. 4.4 the fuzzy optimal solution of (5.16) can be obtained by solving the problem (5.30):

$$\text{Maximize/Minimize} \; \Re(\sum_{j=1}^{n} (p_j, q_j, \alpha'_j, \beta'_j)_{LR} \otimes (x_j, y_j, \alpha''_j, \beta''_j)_{LR})$$

subject to

$$\Re(\sum_{j=1}^{n} (m_{pj}, n_{pj}, \gamma'_{pj}, \delta'_{pj})_{LR}) \leq \Re(b_p, g_p, \gamma_p, \delta_p)_{LR} \quad \forall p \in N_1$$

$$\Re(\sum_{j=1}^{n} (m_{qj}, n_{qj}, \gamma'_{qj}, \delta'_{qj})_{LR}) \geq \Re(b_q, g_q, \gamma_q, \delta_q)_{LR} \quad \forall q \in N_2 \qquad (5.30)$$

$$\sum_{j=1}^{n} m_{rj} = b_r \; \forall r \in N_3, \qquad\qquad \sum_{j=1}^{n} n_{rj} = g_r \; \forall r \in N_3$$

$$\sum_{j=1}^{n} \gamma'_{rj} = \gamma_r \; \forall r \in N_3, \qquad\qquad \sum_{j=1}^{n} \delta'_{rj} = \delta_r \; \forall r \in N_3$$

$$x_j \leq y_j, \alpha''_j \geq 0, \beta''_j \geq 0 \; \forall j = 1, 2, \ldots, n$$

**Step 5** Assuming $(p_j, q_j, \alpha'_j, \beta'_j)_{LR} \otimes (x_j, y_j, \alpha''_j, \beta''_j)_{LR} = (p'_j, q'_j, \alpha'''_j, \beta'''_j)_{LR}$ the problem (5.30) can be written as:

Maximize/Minimize $\Re(\sum\limits_{j=1}^{n}(p'_j, q'_j, \alpha'''_j, \beta'''_j)_{LR})$

subject to

$$\Re(\sum_{j=1}^{n}(m_{pj}, n_{pj}, \gamma'_{pj}, \delta'_{pj})_{LR}) \le \Re(b_p, g_p, \gamma_p, \delta_p)_{LR} \quad \forall\, p \in N_1$$

$$\Re(\sum_{j=1}^{n}(m_{qj}, n_{qj}, \gamma'_{qj}, \delta'_{qj})_{LR}) \ge \Re(b_q, g_q, \gamma_q, \delta_q)_{LR} \quad \forall\, q \in N_2 \qquad (5.31)$$

$$\sum_{j=1}^{n} m_{rj} = b_r \ \ \forall\, r \in N_3, \qquad\qquad \sum_{j=1}^{n} n_{rj} = g_r \ \ \forall\, r \in N_3$$

$$\sum_{j=1}^{n} \gamma'_{rj} = \gamma_r \ \ \forall\, r \in N_3, \qquad\qquad \sum_{j=1}^{n} \delta'_{rj} = \delta_r \ \ \forall\, r \in N_3$$

$$x_j \le y_j, \alpha''_j \ge 0, \beta''_j \ge 0 \ \ \forall\, j = 1, 2, \ldots, n$$

**Step 6** Using the linearity property $\Re(\sum\limits_{j=1}^{n} \tilde{A}_i) = \sum\limits_{j=1}^{n} \Re(\tilde{A}_i)$, where $\tilde{A}_i$ is a fuzzy number, problem (5.31) can be converted into (5.32):

Maximize/Minimize $\sum\limits_{j=1}^{n} \Re(p'_j, q'_j, \alpha'''_j, \beta'''_j)_{LR}$

subject to

$$\Re(\sum_{j=1}^{n}(m_{pj}, n_{pj}, \gamma'_{pj}, \delta'_{pj})_{LR}) \le \Re(b_p, g_p, \gamma_p, \delta_p)_{LR} \quad \forall\, p \in N_1$$

$$\Re(\sum_{j=1}^{n}(m_{qj}, n_{qj}, \gamma'_{qj}, \delta'_{qj})_{LR}) \ge \Re(b_q, g_q, \gamma_q, \delta_q)_{LR} \quad \forall\, q \in N_2 \qquad (5.32)$$

$$\sum_{j=1}^{n} m_{rj} = b_r \ \ \forall\, r \in N_3, \qquad\qquad \sum_{j=1}^{n} n_{rj} = g_r \ \ \forall\, r \in N_3$$

$$\sum_{j=1}^{n} \gamma'_{rj} = \gamma_r \ \ \forall\, r \in N_3, \qquad\qquad \sum_{j=1}^{n} \delta'_{rj} = \delta_r \ \ \forall\, r \in N_3$$

$$x_j \le y_j, \alpha''_j \ge 0, \beta''_j \ge 0 \ \ \forall\, j = 1, 2, \ldots, n$$

**Step 7** Using $\Re(m, n, \alpha, \beta) = \frac{1}{2}(\int_0^1 (m - \alpha L^{-1}(\lambda))\, d\lambda + \int_0^1 (n + \beta R^{-1}(\lambda))\, d\lambda)$ problem (5.32) can be converted into (5.33):

Maximize/Minimize $\sum_{j=1}^{n} (\frac{1}{2}(\int_0^1 (p'_j - \alpha'''_j L^{-1}(\lambda))d\lambda + \int_0^1 (q'_j + \beta'''_j R^{-1}(\lambda))d\lambda))$

subject to

$$\sum_{j=1}^{n} \frac{1}{2}(\int_0^1 (m'_{pj} - \gamma'_{pj}L^{-1}(\lambda))d\lambda + \int_0^1 (n_{pj} + \delta'_{pj}R^{-1}(\lambda))d\lambda)$$

$$\leq \frac{1}{2}(\int_0^1 (b_p - \gamma_p L^{-1}(\lambda))d\lambda + \int_0^1 (g_p + \delta_p R^{-1}(\lambda))d\lambda) \ \forall p \in N_1$$

$$\sum_{j=1}^{n} \frac{1}{2}(\int_0^1 (m'_{qj} - \gamma'_{qj}L^{-1}(\lambda))d\lambda + \int_0^1 (n_{qj} + \delta'_{qj}R^{-1}(\lambda))d\lambda) \qquad (5.33)$$

$$\geq \frac{1}{2}(\int_0^1 (b_q - \gamma_q L^{-1}(\lambda))d\lambda + \int_0^1 (g_q + \delta_q R^{-1}(\lambda))d\lambda) \ \forall q \in N_2$$

$$\sum_{j=1}^{n} m_{rj} = b_r \ \forall r \in N_3, \sum_{j=1}^{n} n_{rj} = g_r \ \forall r \in N_3$$

$$\sum_{j=1}^{n} \gamma'_{rj} = \gamma_r \ \forall r \in N_3, \sum_{j=1}^{n} \delta'_{rj} = \delta_r \ \forall r \in N_3$$

$$x_j \leq y_j, \alpha''_j \geq 0, \beta''_j \geq 0 \ \forall j = 1, 2, \ldots, n$$

**Step 8** Solve the crisp non-linear programming problem (5.33) by using an appropriate existing method [9] to find the optimal solution $\{x^*_j, y^*_j, \alpha''^*_j, \beta''^*_j\}$.

**Step 9** Find the fuzzy optimal solution $\{\tilde{x}^*_j\}$ of the fully fuzzy linear programming problem (5.16) by putting the values of $x^*_j, y^*_j, \alpha''^*_j$ and $\beta''^*_j$ in $\tilde{x}^*_j = (x^*_j, y^*_j, \alpha''^*_j, \beta''^*_j)$.

**Step 10** Find the fuzzy optimal value by putting the values of $\tilde{x}^*_j$, obtained from Step 9, in $\sum_{j=1}^{n} \tilde{c}_j \otimes \tilde{x}^*_j$.

### 5.4.3  Verification of the Presented Methods

In this section, it is depicted that the methods, presented in Sects. 5.4.1 and 5.4.2, are equivalent and the solution obtained by using any of these two methods will be same.

If $\tilde{A}$ and $\tilde{B}$ are any two fuzzy numbers such that $\tilde{A} = \tilde{B}$ then $\Re(\tilde{A}) = \Re(\tilde{B})$ so the 1st and 2nd constraints i.e., $\sum_{j=1}^{n} \tilde{a}_{pj} \otimes \tilde{x}_j \oplus \tilde{S}_p = \tilde{b}_p \oplus \tilde{S}'_p$ and $\sum_{j=1}^{n} \tilde{a}_{qj} \otimes \tilde{x}_j \oplus \tilde{S}_q = \tilde{b}_q \oplus \tilde{S}'_q$ of problem (5.19) can be written as:

$$\Re(\sum_{j=1}^{n} \tilde{a}_{pj} \otimes \tilde{x}_j \oplus \tilde{S}_p) = \Re(\tilde{b}_p \oplus \tilde{S}'_p) \ \forall p \in N_1 \tag{5.34}$$

$$\Re(\sum_{j=1}^{n} \tilde{a}_{qj} \otimes \tilde{x}_j \oplus \tilde{S}_q) = \Re(\tilde{b}_q \oplus \tilde{S}'_q) \ \forall q \in N_2 \tag{5.35}$$

Since, the existing ranking function [3] $\Re$ satisfies the linearity property, so Eqs. (5.34) and (5.35) can be written as:

$$\Re(\sum_{j=1}^{n} \tilde{a}_{pj} \otimes \tilde{x}_j) + \Re(\tilde{S}_p) = \Re(\tilde{b}_p) + \Re(\tilde{S}'_p) \ \forall p \in N_1 \tag{5.36}$$

$$\Re(\sum_{j=1}^{n} \tilde{a}_{qj} \otimes \tilde{x}_j) + \Re(\tilde{S}_q) = \Re(\tilde{b}_q) + \Re(\tilde{S}'_q) \ \forall q \in N_2 \tag{5.37}$$

Equations (5.36) and (5.37) can be written as:

$$\Re(\tilde{b}_p) - \Re(\sum_{j=1}^{n} \tilde{a}_{pj} \otimes \tilde{x}_j) = \Re(\tilde{S}_p) - \Re(\tilde{S}'_p) \ \forall p \in N_1 \tag{5.38}$$

$$\Re(\tilde{b}_q) - \Re(\sum_{j=1}^{n} \tilde{a}_{qj} \otimes \tilde{x}_j) = \Re(\tilde{S}_q) - \Re(\tilde{S}'_q) \ \forall q \in N_2 \tag{5.39}$$

Using 4th and 5th constraint of problem (5.19) i.e., $\Re(\tilde{S}_p) - \Re(\tilde{S}'_p) \geq 0$ and $\Re(\tilde{S}_q) - \Re(\tilde{S}'_q) \leq 0$ Eqs. (5.38) and (5.39) can be written as:

$$\Re(\tilde{b}_p) - \Re(\sum_{j=1}^{n} \tilde{a}_{pj} \otimes \tilde{x}_j) \geq 0 \ \forall p \in N_1$$

$$\Re(\tilde{b}_q) - \Re(\sum_{j=1}^{n} \tilde{a}_{qj} \otimes \tilde{x}_j) \leq 0 \ \forall q \in N_2$$

i.e.,

$$\Re(\sum_{j=1}^{n} \tilde{a}_{pj} \otimes \tilde{x}_j) \leq \Re(\tilde{b}_p) \ \forall p \in N_1$$

$$\Re(\sum_{j=1}^{n} \tilde{a}_{qj} \otimes \tilde{x}_j) \geq \Re(\tilde{b}_q) \ \forall q \in N_2$$

So, the methods, presented in Sects. 5.4.1 and 5.4.2 are equivalent hence the solution obtained by using any of these two methods will be same.

## 5.5    Illustrative Example

In this section, the methods presented in this chapter are illustrated with the help of fully fuzzy linear programming problem, chosen in Example 5.10, which cannot be solved by using the existing method [2].

### 5.5.1    Fuzzy Optimal Solution of the Chosen Problem by Using the Kumar and Kaur's Method

Using the method, presented in Sect. 5.4.1, the fuzzy optimal solution of the fully fuzzy linear programming problem, chosen in Example 5.10, can be obtained as follows:

**Step 1** Using Step 2 of the method, presented in Sect. 5.4.1, the fully fuzzy linear programming problem, chosen in Example 5.10, can be written as:

Maximize $((4, 4, 0, 0)_{LR} \otimes \tilde{x}_1 \oplus (1, 1, 1, 1)_{LR} \otimes \tilde{x}_2)$

subject to

$(2, 5, 5, 2)_{LR} \otimes \tilde{x}_1 \oplus (-1, 5, 1, 2)_{LR} \otimes \tilde{x}_2 \oplus \tilde{S} = (-17, 45, 25, 46)_{LR} \oplus \tilde{S}'$

$(1, 2, 1, 1)_{LR} \otimes \tilde{x}_1 \oplus (3, 5, 2, 2)_{LR} \otimes \tilde{x}_2 = (7, 24, 6, 24)_{LR}$

where $\tilde{x}_1, \tilde{x}_2, \tilde{S}, \tilde{S}'$ are $LR$ flat fuzzy numbers and $L(x) = \max\{0, 1 - x\}$, $R(x) = \max\{0, 1 - x^2\}$.

**Step 2** Assuming $\tilde{x}_1 = (x_1, y_1, \alpha_1, \beta_1)_{LR}$, $\tilde{x}_2 = (x_2, y_2, \alpha_2, \beta_2)_{LR}$, $\tilde{S} = (s, t, \eta, \rho)_{LR}$ and $\tilde{S}' = (s', t', \eta', \rho')_{LR}$ the fully fuzzy linear programming problem, obtained in Step 1, can be written as:

Maximize$((4, 4, 0, 0)_{LR} \otimes (x_1, y_1, \alpha_1, \beta_1)_{LR} \oplus (1, 1, 1, 1)_{LR} \otimes (x_2, y_2, \alpha_2, \beta_2)_{LR})$

subject to

$(2, 5, 5, 2)_{LR} \otimes (x_1, y_1, \alpha_1, \beta_1)_{LR} \oplus (-1, 5, 1, 2)_{LR} \otimes (x_2, y_2, \alpha_2, \beta_2)_{LR}$

$\oplus (s, t, \eta, \rho)_{LR} = (-17, 45, 25, 46)_{LR} \oplus (s', t', \eta', \rho')_{LR}$

$(1, 2, 1, 1)_{LR} \otimes (x_1, y_1, \alpha_1, \beta_1)_{LR} \oplus (3, 5, 2, 2)_{LR} \otimes (x_2, y_2, \alpha_2, \beta_2)_{LR}$

$= (7, 24, 6, 24)_{LR}$

where $(x_1, y_1, \alpha_1, \beta_1)_{LR}$, $(x_2, y_2, \alpha_2, \beta_2)_{LR}$, $(s, t, \eta, \rho)_{LR}$ and $(s', t', \eta', \rho')_{LR}$ are $LR$ flat fuzzy numbers.

**Step 3** Using the product, presented in Sect. 4.2.1, the fully fuzzy linear programming problem, obtained in Step 2, can be written as:

Maximize $((\min\{4x_1, 4x_1\}, \max\{4y_1, 4y_1\}, \min\{4x_1, 4x_1\} - \min\{4x_1 - 4\alpha_1, 4x_1 - 4\alpha_1\}, \max\{4y_1 + 4\beta_1, 4y_1 + 4\beta_1\} - \max\{4y_1, 4y_1\})_{LR} \oplus (\min\{x_2, x_2\}, \max\{y_2, y_2\}, \min\{x_2, x_2\} - \min\{0, 2x_2 - 2\alpha_2\}, \max\{0, 2y_2 + 2\beta_2\} - \max\{y_2, y_2\})_{LR})$

subject to

$(\min\{2x_1, 5x_1\}, \max\{2y_1, 5y_1\}, \min\{2x_1, 5x_1\} - \min\{-3y_1 - 3\beta_1, 7x_1 - 7\alpha_1\}, \max\{7y_1 + 7\beta_1, -3x_1 + 3\alpha_1\} - \max\{2y_1, 5y_1\})_{LR} \oplus (\min\{-y_2, 5x_2\}, \max\{-x_2, 5y_2\}, \min\{-y_2, 5x_2\} - \min\{-2y_2 - 2\beta_2, 7x_2 - 7\alpha_2\}, \max\{-2x_2 + 2\alpha_2, 7y_2 + 7\beta_2\} - \max\{-x_2, 5y_2\})_{LR} \oplus (s, t, \eta, \rho)_{LR} = (-17, 45, 25, 46)_{LR} \oplus (s', t', \eta', \rho')_{LR}$ $(\min\{x_1, 2x_1\}, \max\{y_1, 2y_1\}, \min\{x_1, 2x_1\} - \min\{0, 3x_1 - 3\alpha_1\}, \max\{0, 3y_1 + 3\beta_1\} - \max\{y_1, 2y_1\})_{LR} \oplus (\min\{3x_2, 5x_2\}, \max\{3y_2, 5y_2\}, \min\{3x_2, 5x_2\} - \min\{x_2 - \alpha_2, 7x_2 - 7\alpha_2\}, \max\{y_2 + \beta_2, 7y_2 + 7\beta_2\} - \max\{3y_2, 5y_2\})_{LR} = (7, 24, 6, 24)_{LR}$

where $(x_1, y_1, \alpha_1, \beta_1)_{LR}, (x_2, y_2, \alpha_2, \beta_2)_{LR}, (s, t, \eta, \rho)_{LR}$ and $(s', t', \eta', \rho')_{LR}$ are $LR$ flat fuzzy numbers.

**Step 4** Using the arithmetic operations, defined in Sect. 4.1.2 and Definition 4.5, the fully fuzzy linear programming problem, obtained in Step 3, can be written as:

Maximize $((\min\{4x_1, 4x_1\} + \min\{x_2, x_2\}, \max\{4y_1, 4y_1\} + \max\{y_2, y_2\}, \min\{4x_1, 4x_1\} - \min\{4x_1 - 4\alpha_1, 4x_1 - 4\alpha_1\} + \min\{x_2, x_2\} - \min\{0, 2x_2 - 2\alpha_2\}, \max\{4y_1 + 4\beta_1, 4y_1 + 4\beta_1\} - \max\{4y_1, 4y_1\} + \max\{0, 2y_2 + 2\beta_2\} - \max\{y_2, y_2\})_{LR})$

subject to

$$\min\{2x_1, 5x_1\} + \min\{-y_2, 5x_2\} + s = -17 + s'$$
$$\max\{2y_1, 5y_1\} + \max\{-x_2, 5y_2\} + t = 45 + t'$$
$$\min\{x_1, 2x_1\} + \min\{3x_2, 5x_2\} = 7$$
$$\max\{y_1, 2y_1\} + \max\{3y_2, 5y_2\} = 24$$

$\min\{2x_1, 5x_1\} - \min\{-3y_1 - 3\beta_1, 7x_1 - 7\alpha_1\} + \min\{-y_2, 5x_2\} - \min\{-2y_2 - 2\beta_2, 7x_2 - 7\alpha_2\} + \eta = 25 + \eta'$

$\max\{7y_1 + 7\beta_1, -3x_1 + 3\alpha_1\} - \max\{2y_1, 5y_1\} + \max\{-2x_2 + 2\alpha_2, 7y_2 + 7\beta_2\} - \max\{-x_2, 5y_2\} + \rho = 46 + \rho'$

$\min\{x_1, 2x_1\} - \min\{0, 3x_1 - 3\alpha_1\} + \min\{3x_2, 5x_2\} - \min\{x_2 - \alpha_2, 7x_2 - 7\alpha_2\} = 6$

$\max\{0, 3y_1 + 3\beta_1\} - \max\{y_1, 2y_1\} + \max\{y_2 + \beta_2, 7y_2 + 7\beta_2\} - \max\{3y_2, 5y_2\} = 24$

$x_1 \leq y_1, \alpha_1 \geq 0, \beta_1 \geq 0, x_2 \leq y_2, \alpha_2 \geq 0, \beta_2 \geq 0$

$s \leq t, \eta \geq 0, \rho \geq 0, s' \leq t', \eta' \geq 0, \rho' \geq 0$

**Step 5** Using Step 7 of the method, presented in Sect. 5.4.1, the fuzzy linear programming problem, obtained in Step 4, can be written as:

Maximize $\Re((\min\{4x_1, 4x_1\} + \min\{x_2, x_2\}, \max\{4y_1, 4y_1\} + \max\{y_2, y_2\}, \min$
$\{4x_1, 4x_1\} - \min\{4x_1 - 4\alpha_1, 4x_1 - 4\alpha_1\} + \min\{x_2, x_2\} - \min\{0, 2x_2 - 2\alpha_2\}, \max$
$\{4y_1 + 4\beta_1, 4y_1 + 4\beta_1\} - \max\{4y_1, 4y_1\} + \max\{0, 2y_2 + 2\beta_2\} - \max\{y_2, y_2\})_{LR})$
subject to

$$\min\{2x_1, 5x_1\} + \min\{-y_2, 5x_2\} + s = -17 + s'$$
$$\max\{2y_1, 5y_1\} + \max\{-x_2, 5y_2\} + t = 45 + t'$$
$$\min\{x_1, 2x_1\} + \min\{3x_2, 5x_2\} = 7$$
$$\max\{y_1, 2y_1\} + \max\{3y_2, 5y_2\} = 24$$

$\min\{2x_1, 5x_1\} - \min\{-3y_1 - 3\beta_1, 7x_1 - 7\alpha_1\} + \min\{-y_2, 5x_2\} - \min\{-2y_2 - 2\beta_2, 7x_2 - 7\alpha_2\} + \eta = 25 + \eta'$
$\max\{7y_1 + 7\beta_1, -3x_1 + 3\alpha_1\} - \max\{2y_1, 5y_1\} + \max\{-2x_2 + 2\alpha_2, 7y_2 + 7\beta_2\} - \max\{-x_2, 5y_2\} + \rho = 46 + \rho'$
$\min\{x_1, 2x_1\} - \min\{0, 3x_1 - 3\alpha_1\} + \min\{3x_2, 5x_2\} - \min\{x_2 - \alpha_2, 7x_2 - 7\alpha_2\}$
$= 6$
$\max\{0, 3y_1 + 3\beta_1\} - \max\{y_1, 2y_1\} + \max\{y_2 + \beta_2, 7y_2 + 7\beta_2\} - \max\{3y_2, 5y_2\}$
$= 24$
$x_1 \leq y_1, \alpha_1 \geq 0, \beta_1 \geq 0, x_2 \leq y_2, \alpha_2 \geq 0, \beta_2 \geq 0$
$s \leq t, \eta \geq 0, \rho \geq 0, s' \leq t', \eta' \geq 0, \rho \geq 0$

**Step 6** Using Steps 9 and 10 of the method, presented in Sect. 5.4.1, the fuzzy linear programming problem, obtained in Step 5, can be written as:

Maximize $\frac{1}{3}|y_2 + \beta_2| - \frac{1}{4}|x_2 - \alpha_2| - \alpha_1 + 2x_1 - \frac{1}{4}\alpha_2 + \frac{1}{2}x_2 + 2y_1 + \frac{4}{3}\beta_1$
$+ \frac{1}{2}y_2 + \frac{1}{3}\beta_2$
subject to

$$\frac{7}{2}x_1 - \frac{3}{2}|x_1| - \frac{1}{2}y_2 + \frac{5}{2}x_2 - \frac{1}{2}|y_2 + 5x_2| + s = -17 + s'$$
$$\frac{7}{2}y_1 + \frac{3}{2}|y_1| - \frac{1}{2}x_2 + \frac{5}{2}y_2 + \frac{1}{2}|x_2 + 5y_2| + t = 45 + t'$$
$$\frac{3}{2}x_1 - \frac{1}{2}|x_1| + 4x_2 - |x_2| = 7$$
$$\frac{3}{2}y_1 + \frac{1}{2}|y_1|4y_2 + |y_2| = 24$$

$$-\frac{3}{2}|x_1| + \frac{3}{2}y_1 + \frac{3}{2}\beta_1 + \frac{7}{2}\alpha_1 + \frac{1}{2}|3y_1 + 3\beta_1 + 7x_1 - 7\alpha_1| + \frac{1}{2}y_2 - x_2 -$$

$$\frac{1}{2}|y_2 + 5x_2| + \beta_2 + \frac{7}{2}\alpha_2 + \frac{1}{2}|2y_2 + 2\beta_2 + 7x_2 - 7\alpha_2| + \eta = 25 + \eta'$$

$$\frac{7}{2}\beta_1 - \frac{3}{2}x_1 + \frac{3}{2}\alpha_1 + \frac{1}{2}|7y_1 + 7\beta_1 + 3x_1 - 3\alpha_1| - \frac{3}{2}|y_1| - \frac{1}{2}x_2 + \alpha_2 + y_2$$

$$+\frac{7}{2}\beta_2 + \frac{1}{2}|7y_2 + 7\beta_2 + 2x_2 - 2\alpha_2| - \frac{1}{2}|x_2 + 5y_2| + \rho = 46 + \rho'$$

$$-\frac{1}{2}|x_1| + \frac{3}{2}\alpha_1 + \frac{3}{2}|x_1 - \alpha_1| - |x_2| + 4\alpha_2 + 3|x_2 - \alpha_2| = 6$$

$$\frac{3}{2}\beta_1 + \frac{3}{2}|y_1 + \beta_1| - \frac{1}{2}|y_1| + 4\beta_2 + 3|y_2 + \beta_2| - |y_2| = 24$$

$$x_1 \le y_1, \alpha_1 \ge 0, \beta_1 \ge 0, x_2 \le y_2, \alpha_2 \ge 0, \beta_2 \ge 0$$

$$s \le t, \eta \ge 0, \rho \ge 0, s' \le t', \eta' \ge 0, \rho \ge 0$$

**Step 7** The optimal solution of the crisp non-linear programming problem, obtained in Step 6, is $x_1 = 4, y_1 = 4, \alpha_1 = 0, \beta_1 = \frac{51}{335}, x_2 = 1, y_2 = \frac{16}{5}, \alpha_2 = 0$ and $\beta_2 = \frac{629}{335}$.

**Step 8** Putting the values of $x_1, y_1, \alpha_1, \beta_1, x_2, y_2, \alpha_2$ and $\beta_2$ in $\tilde{x}_1 = (x_1, y_1, \alpha_1, \beta_1)$ and $\tilde{x}_2 = (x_2, y_2, \alpha_2, \beta_2)$, the exact fuzzy optimal solution is $\tilde{x}_1 = (4, 4, 0, \frac{51}{335})_{LR}, \tilde{x}_2 = (1, \frac{16}{5}, 0, \frac{629}{335})_{LR}$.

**Step 9** Putting the values of $\tilde{x}_1$ and $\tilde{x}_2$, obtained from Step 8, in the objective function the fuzzy optimal value is $(17, \frac{96}{5}, 1, \frac{2534}{335})_{LR}$.

### 5.5.2 Fuzzy Optimal Solution of the Chosen Problem by Using the Alternative Method

Using the method, presented in Sect. 5.4.2, the fuzzy optimal solution of the fully fuzzy linear programming problem, chosen in Example 5.10, can be obtained as follows:

**Step 1** Assuming $\tilde{x}_1 = (x_1, y_1, \alpha_1, \beta_1)_{LR}$ and $\tilde{x}_2 = (x_2, y_2, \alpha_2, \beta_2)_{LR}$ the fully fuzzy linear programming problem, chosen in Example 5.10, can be written as:

Maximize $((4, 4, 0, 0)_{LR} \otimes (x_1, y_1, \alpha_1, \beta_1)_{LR} \oplus (1, 1, 1, 1)_{LR} \otimes (x_2, y_2, \alpha_2, \beta_2)_{LR})$

subject to

$(2, 5, 5, 2)_{LR} \otimes (x_1, y_1, \alpha_1, \beta_1)_{LR} \oplus (-1, 5, 1, 2)_{LR} \otimes (x_2, y_2, \alpha_2, \beta_2)_{LR}$

$\preceq (-17, 45, 25, 46)_{LR}$

$(1, 2, 1, 1)_{LR} \otimes (x_1, y_1, \alpha_1, \beta_1)_{LR} \oplus (3, 5, 2, 2)_{LR} \otimes (x_2, y_2, \alpha_2, \beta_2)_{LR}$

$= (7, 24, 6, 24)_{LR}$

where $(x_1, y_1, \alpha_1, \beta_1)_{LR}, (x_2, y_2, \alpha_2, \beta_2)_{LR}$ are $LR$ flat fuzzy numbers and $L(x) = \max\{0, 1 - x\}, R(x) = \max\{0, 1 - x^2\}$.

**Step 2** Using the product, presented in Sect. 4.2.1, the fully fuzzy linear programming problem, obtained in Step 1, can be written as:

Maximize $((\min\{4x_1, 4x_1\}, \max\{4y_1, 4y_1\}, \min\{4x_1, 4x_1\} - \min\{4x_1 - 4\alpha_1,$

$4x_1 - 4\alpha_1\}, \max\{4y_1 + 4\beta_1, 4y_1 + 4\beta_1\} - \max\{4y_1, 4y_1\})_{LR} \oplus (\min\{x_2, x_2\}, \max$

$\{y_2, y_2\}, \min\{x_2, x_2\} - \min\{0, 2x_2 - 2\alpha_2\}, \max\{0, 2y_2 + 2\beta_2\} - \max\{y_2, y_2\})_{LR})$

subject to

$(\min\{2x_1, 5x_1\}, \max\{2y_1, 5y_1\}, \min\{2x_1, 5x_1\} - \min\{-3y_1 - 3\beta_1, 7x_1 - 7\alpha_1\},$

$\max\{7y_1 + 7\beta_1, -3x_1 + 3\alpha_1\} - \max\{2y_1, 5y_1\})_{LR} \oplus (\min\{-y_2, 5x_2\}, \max\{-x_2,$

$5y_2\}, \min\{-y_2, 5x_2\} - \min\{-2y_2 - 2\beta_2, 7x_2 - 7\alpha_2\}, \max\{-2x_2 + 2\alpha_2, 7y_2 + 7\beta_2\}$

$- \max\{-x_2, 5y_2\})_{LR} \preceq (-17, 45, 25, 46)_{LR}$

$(\min\{x_1, 2x_1\}, \max\{y_1, 2y_1\}, \min\{x_1, 2x_1\} - \min\{0, 3x_1 - 3\alpha_1\}, \max\{0, 3y_1 + 3\beta_1\}$

$- \max\{y_1, 2y_1\})_{LR} \oplus (\min\{3x_2, 5x_2\}, \max\{3y_2, 5y_2\}, \min\{3x_2, 5x_2\} - \min\{x_2 - \alpha_2,$

$7x_2 - 7\alpha_2\}, \max\{y_2 + \beta_2, 7y_2 + 7\beta_2\} - \max\{3y_2, 5y_2\})_{LR} = (7, 24, 6, 24)_{LR}$

where $(x_1, y_1, \alpha_1, \beta_1)_{LR}$ and $(x_2, y_2, \alpha_2, \beta_2)_{LR}$ are $LR$ flat fuzzy numbers.

**Step 3** Using Step 4 of the method, presented in Sect. 5.4.2, the fully fuzzy linear programming problem, obtained in Step 2, can be written as:

Maximize $\Re((\min\{4x_1, 4x_1\}, \max\{4y_1, 4y_1\}, \min\{4x_1, 4x_1\} - \min\{4x_1 - 4\alpha_1, 4x_1$

$- 4\alpha_1\}, \max\{4y_1 + 4\beta_1, 4y_1 + 4\beta_1\} - \max\{4y_1, 4y_1\})_{LR} \oplus (\min\{x_2, x_2\}, \max$

$\{y_2, y_2\}, \min\{x_2, x_2\} - \min\{0, 2x_2 - 2\alpha_2\}, \max\{0, 2y_2 + 2\beta_2\} - \max\{y_2, y_2\})_{LR})$

subject to

$\Re((\min\{2x_1, 5x_1\}, \max\{2y_1, 5y_1\}, \min\{2x_1, 5x_1\} - \min\{-3y_1 - 3\beta_1, 7x_1 - 7\alpha_1\},$

$\max\{7y_1 + 7\beta_1, -3x_1 + 3\alpha_1\} - \max\{2y_1, 5y_1\})_{LR} \oplus (\min\{-y_2, 5x_2\}, \max\{-x_2,$

$5y_2\}, \min\{-y_2, 5x_2\} - \min\{-2y_2 - 2\beta_2, 7x_2 - 7\alpha_2\}, \max\{-2x_2 + 2\alpha_2, 7y_2 +$

$7\beta_2\} - \max\{-x_2, 5y_2\})_{LR}) \leq \Re(-17, 45, 25, 46)_{LR}$

$$\min\{x_1, 2x_1\} + \min\{3x_2, 5x_2\} = 7$$
$$\max\{y_1, 2y_1\} + \max\{3y_2, 5y_2\} = 24$$

$$\min\{x_1, 2x_1\} - \min\{0, 3x_1 - 3\alpha_1\} + \min\{3x_2, 5x_2\} - \min\{x_2 - \alpha_2, 7x_2 - 7\alpha_2\}$$
$$= 6$$
$$\max\{0, 3y_1 + 3\beta_1\} - \max\{y_1, 2y_1\} + \max\{y_2 + \beta_2, 7y_2 + 7\beta_2\} - \max\{3y_2, 5y_2\})_{LR}$$
$$= 24$$
$$x_1 \le y_1, \alpha_1 \ge 0, \beta_1 \ge 0, x_2 \le y_2, \alpha_2 \ge 0, \beta_2 \ge 0$$

**Step 4** Using Steps 6 and 7 of the method, presented in Sect. 5.4.2, the fully fuzzy linear programming problem, obtained in Step 3, can be written as:

$$\text{Maximize } \frac{1}{3}|y_2 + \beta_2| - \frac{1}{4}|x_2 - \alpha_2| - \alpha_1 + 2x_1 - \frac{1}{4}\alpha_2 + \frac{1}{2}x_2 + 2y_1 + \frac{4}{3}\beta_1$$
$$+ \frac{1}{2}y_2 + \frac{1}{3}\beta_2$$

subject to

$$-\frac{3}{2}|5x_2 + y_2| + |x_2 + 5y_2| - \frac{3}{2}| - 7x_1 + 7\alpha_1 - 3y_1 - 3\beta_1| + 2| - 3x_1 + 3\alpha_1$$

$$- 7y_1 - 7\beta_1| - \frac{9}{2}|x_1| + 3|y_1| + 2|2\alpha_2 - 2x_2 - 7y_2 - 7\beta_2| - \frac{3}{2}|7\alpha_2 - 7x_2 - 2y_2$$

$$- 2\beta_2| + 15x_1 - \frac{9}{2}\alpha_1 - \frac{13}{2}\alpha_2 + 13x_2 + \frac{33}{2}y_1 + \frac{19}{2}\beta_1 + \frac{29}{2}y_2 + 11\beta_2 \le 277$$

$$\frac{3}{2}x_1 - \frac{1}{2}|x_1| + 4x_2 - |x_2| = 7$$

$$\frac{3}{2}y_1 + \frac{1}{2}|y_1|4y_2 + |y_2| = 24$$

$$-\frac{1}{2}|x_1| + \frac{3}{2}\alpha_1 + \frac{3}{2}|x_1 - \alpha_1| - |x_2| + 4\alpha_2 + 3|x_2 - \alpha_2| = 6$$

$$\frac{3}{2}\beta_1 + \frac{3}{2}|y_1 + \beta_1| - \frac{1}{2}|y_1| + 4\beta_2 + 3|y_2 + \beta_2| - |y_2| = 24$$

$$x_1 \le y_1, \alpha_1 \ge 0, \beta_1 \ge 0, x_2 \le y_2, \alpha_2 \ge 0, \beta_2 \ge 0$$

**Step 5** The optimal solution of the crisp non-linear programming problem, obtained in Step 4, is $x_1 = 4, y_1 = 4, \alpha_1 = 0, \beta_1 = \frac{51}{335}, x_2 = 1, y_2 = \frac{16}{5}, \alpha_2 = 0$ and $\beta_2 = \frac{629}{335}$.

**Step 6** Putting the values of $x_1, y_1, \alpha_1, \beta_1, x_2, y_2, \alpha_2$ and $\beta_2$ in $\tilde{x}_1 = (x_1, y_1, \alpha_1, \beta_1)$ and $\tilde{x}_2 = (x_2, y_2, \alpha_2, \beta_2)$, the exact fuzzy optimal solution is $\tilde{x}_1 = (4, 4, 0, \frac{51}{335})_{LR}, \tilde{x}_2 = (1, \frac{16}{5}, 0, \frac{629}{335})_{LR}$.

**Step 7** Putting the values of $\tilde{x}_1$ and $\tilde{x}_2$, obtained from Step 6, in the objective function, the fuzzy optimal value is $(17, \frac{96}{5}, 1, \frac{2534}{335})_{LR}$.

## 5.6  Advantages of the Presented Methods

In this section, the advantages of the methods, presented in this chapter, over the existing methods [2, 4–8] and the methods, presented in previous chapters are discussed.

The main advantage of the methods, presented in this chapter, is that all the fully fuzzy linear programming problems which can be solved by using the existing methods [2, 4–8] and the methods presented in previous chapters can also be solved by using the methods presented in this chapter. Furthermore, on solving these problems by using the existing methods [2, 4–8], methods presented in previous chapters and the methods presented in this chapter, same results are obtained.

However, there exist several problems which cannot be solved by using the existing methods [2, 4–8] and the methods, presented in previous chapters but can be solved by using the methods, presented in this chapter.

## 5.7  Comparative Study

To compare the existing methods [2, 4–8] and the methods presented in this chapter, the results of fuzzy linear programming problems and fully fuzzy linear programming problems, obtained by using the existing methods and the methods presented in this chapter, are shown in Table 5.1.

It is obvious from the results, shown in Table 5.1, that the fuzzy linear programming problems and fully fuzzy linear programming problems, chosen in Examples 5.6–5.11, cannot be solved by using any of the existing methods but all these problems can be solved by using the methods, presented in this chapter. Also, all the fuzzy linear programming problems and fully fuzzy linear programming problems, chosen in Examples 5.1–5.5, for which there is a need to apply the different existing methods can be easily solved by using the methods, presented in this chapter i.e., all the fuzzy linear programming problems and fully fuzzy linear programming problems which may or may not be solved by using the existing methods can be solved by using the methods, presented in this chapter.

**Table 5.1** Results of chosen problems by existing methods and the methods presented in this chapter

| Example | Fuzzy optimal value | | | | | | |
|---|---|---|---|---|---|---|---|
| | Existing method [7] | Existing method [5] | Existing method [8] | Existing method [4] | Existing method [6] | Existing method [2] | Methods presented in this chapter |
| 5.1 | $(\frac{151}{24}, \frac{277}{32}, \frac{227}{96}, \frac{377}{96})$ | $(\frac{151}{24}, \frac{277}{32}, \frac{227}{96}, \frac{377}{96})$ | Not applicable | Not applicable | Not applicable | $(\frac{151}{24}, \frac{277}{32}, \frac{227}{96}, \frac{377}{96})$ | $(\frac{151}{24}, \frac{277}{32}, \frac{227}{96}, \frac{377}{96})$ |
| 5.2 | $(\frac{90}{7}, \frac{148}{7}, \frac{32}{7}, \frac{90}{7})$ | $(\frac{90}{7}, \frac{148}{7}, \frac{32}{7}, \frac{90}{7})$ | $(\frac{90}{7}, \frac{148}{7}, \frac{32}{7}, \frac{90}{7})$ | Not applicable | Not applicable | $(\frac{90}{7}, \frac{148}{7}, \frac{32}{7}, \frac{90}{7})$ | $(\frac{90}{7}, \frac{148}{7}, \frac{32}{7}, \frac{90}{7})$ |
| 5.3 | Not applicable | Not applicable | Not applicable | $(\frac{94235}{169}, \frac{120265}{169}, \frac{19819}{169}, \frac{19819}{169})$ | Not applicable | $(\frac{8250}{13}, \frac{8250}{13}, 0, 0)$ | $(\frac{8250}{13}, \frac{8250}{13}, 0, 0)$ |
| 5.4 | Not applicable | Not applicable | Not applicable | Not applicable | $(\frac{-62}{7}, \frac{300}{7}, \frac{360}{7}, \frac{418}{7})$ | $(\frac{267}{14}, \frac{267}{14}, 0, 0)$ | $(\frac{267}{14}, \frac{267}{14}, 0, 0)$ |
| 5.5 | Not applicable | Not applicable | Not applicable | Not applicable | Not applicable | $(8.25, 16.5, 79.25)_{LR}$ | Infeasible solution |
| 5.6 | Not applicable | Not applicable | Not applicable | Not applicable | Not applicable | Not applicable | $(\frac{179}{121}, \frac{2037}{242}, \frac{456}{121}, \frac{929}{242})_{LR}$ |
| 5.7 | Not applicable | Not applicable | Not applicable | Not applicable | Not applicable | Not applicable | $(\frac{58}{7}, \frac{148}{7}, \frac{90}{7}, \frac{84}{7})_{LR}$ |
| 5.8 | Not applicable | Not applicable | Not applicable | Not applicable | Not applicable | Not applicable | $(0, 55, 11, 33)_{LR}$ |
| 5.9 | Not applicable | Not applicable | Not applicable | Not applicable | Not applicable | Not applicable | $(\frac{168}{9}, \frac{168}{9}, 0, 0)_{LR}$ |
| 5.10 | Not applicable | Not applicable | Not applicable | Not applicable | Not applicable | Not applicable | $(17, \frac{96}{5}, 1, \frac{2534}{335})_{LR}$ |
| 5.11 | Not applicable | Not applicable | Not applicable | Not applicable | Not applicable | Not applicable | $(\frac{2579}{110}, \frac{2579}{110}, \frac{67}{110}, \frac{67}{110})_{LR}$ |

*Remark 5.1* Let $\tilde{A} = (m_1, \alpha_1, \beta_1)_{LR}$ be a non-positive *LR* fuzzy number and $\tilde{B} = (m_2, \alpha_2, \beta_2)_{LR}$ be a non-negative *LR* fuzzy number. Then,

(i) According to the existing product [2],

$$\tilde{A} \otimes \tilde{B} = (m_1 m_2, \alpha_1 m_2 - m_1 \alpha_2, \beta_1 m_2 - m_1 \beta_2)_{LR}$$

(ii) According to the existing product [10],

$$\tilde{A} \otimes \tilde{B} = (m_1 m_2, \alpha_1 m_2 - m_1 \beta_2, \beta_1 m_2 - m_1 \alpha_2)_{LR}$$

Using the product (i) Example 5.5 has a fuzzy optimal solution but by using product (ii) and by using the product, presented in Sect. 4.2.1, Example 5.5 has infeasible solution.

## 5.8　Conclusions

On the basis of the present study, it can be concluded that it is better to use the methods, presented in this chapter as compared to the existing methods [2, 4–8] and methods presented in previous chapters for solving fully fuzzy linear programming problems.

## References

1. Kumar, A., Kaur, J.: Fuzzy optimal solution of fully fuzzy linear programming problems using ranking function. J. Intell. Fuzzy Syst. **26**, 337–344 (2014)
2. Allahviranloo, T., Lotfi, F.H., Kiasary, M.K., Kiani, N.A., Alizadeh, L.: Solving fully fuzzy linear programming problem by the ranking functhion. Appl. Math. Sci. **2**, 19–32 (2008)
3. Yager, R.R.: A procedure for ordering fuzzy subsets of the unit interval. Inf. Sci. **24**, 143–161 (1981)
4. Ganesan, K., Veeramani, P.: Fuzzy linear programs with trapezoidal fuzzy numbers. Ann. Oper. Res. **143**, 305–315 (2006)
5. Mahdavi-Amiri, N., Nasseri, S.H.: Duality in fuzzy number linear programming by the use of a certain linear ranking function. Appl. Math. Comput. **180**, 206–216 (2006)
6. Mahdavi-Amiri, N., Nasseri, S.H.: Duality results and a dual simplex method for linear programming problems with trapezoidal fuzzy variables. Fuzzy Sets Syst. **158**, 1961–1978 (2007)
7. Maleki, H.R., Tata, M., Mashinchi, M.: Linear programming with fuzzy variables. Fuzzy Sets Syst. **109**, 21–33 (2000)
8. Nasseri, S.H., Ardil, E., Yazdani, A., Zaefarian, R.: Simplex method for solving linear programming problems with fuzzy numbers. In: Proceedings of World Academy of Science, Engineering and Technology, vol. 10, pp. 284–288 (2005)
9. Taha, H.A.: Operations Research: An Introduction. Prentice-Hall, New Jersey (2003)
10. Dubois, D., Prade, H.: Fuzzy Sets and Systems, Theory and Applications. Academic Press, New York (1980)

# Chapter 6
# Unique Fuzzy Optimal Value of Fully Fuzzy Linear Programming Problems with Equality Constraints Having LR Flat Fuzzy Numbers

In this chapter, it is shown that the fuzzy optimal value, obtained by using the method, presented in Chap. 4, is not necessarily a unique fuzzy number. So, it does not conform to the uniqueness property of fuzzy optimal value. To overcome this limitation of the method, presented in Chap. 4, a method, proposed by Kaur and Kumar [1, 2], is presented in this chapter for solving fully fuzzy linear programming problems with equality constraints.

## 6.1 Limitations of the Previous Presented Method

In this section, limitations of the method for solving fully fuzzy linear programming problems with equality constraints, presented in Chap. 4, are pointed out.

Let $\{x_j\}$ and $A$ be the optimal solution and optimal value of a linear programming problem respectively. If there exist any feasible solution $\{y_j\}$ of the same linear programming problem such that the value of the objective function of the linear programming problem corresponding to $\{y_j\}$ is also $A$ then $\{y_j\}$ is said to be an alternative optimal solution of the same linear programming problem i.e. corresponding to all alternative optimal solutions the values of objective function should be same.

In this section, a fully fuzzy linear programming problem, chosen in Example 6.1, is solved to show that the results of the fully fuzzy linear programming problem, chosen in Example 6.1, obtained by using the method, presented in Chap. 4, do not conform to the property of alternative optimal solutions.

---

Some of the contents of this chapter are published in *Control and Cybernetics* 41 (2012) 171–182 and the remaining are published in *Journal of Optimization Theory and Applications* 156 (2013) 529–534.

*Example 6.1*

Maximize $((2, 2, 2, 2)_{LR} \otimes \tilde{x}_1 \oplus (7, 12, 2, 2)_{LR} \otimes \tilde{x}_2 \oplus (7, 8, 2, 2)_{LR} \otimes \tilde{x}_3)$

subject to

$\tilde{x}_1 \oplus \tilde{x}_2 = (1, 1, 0, 0)_{LR}$

$\tilde{x}_1 = \tilde{x}_3$

$\tilde{x}_2 \oplus \tilde{x}_3 = (1, 1, 0, 0)_{LR}$

where $\tilde{x}_1, \tilde{x}_2, \tilde{x}_3$ are non-negative $LR$ flat fuzzy numbers and $L(x) = R(x) = \max\{0, 1 - x\}$.

**Solution**: On solving the chosen fully fuzzy linear programming problem by using the method, presented in Chap. 4, it is found that all the fuzzy feasible solutions $\tilde{x}_1 = \tilde{x}_3 = (a, a, 0, 0)_{LR}$ and $\tilde{x}_2 = (1 - a, 1 - a, 0, 0)_{LR}, 0 \leq a \leq 1$ are fuzzy optimal solutions of the chosen fully fuzzy linear programming problem. Putting $\tilde{x}_1 = \tilde{x}_3 = (a, a, 0, 0)_{LR}$ and $\tilde{x}_2 = (1 - a, 1 - a, 0, 0)_{LR}$ in the objective function the fuzzy optimal value is $(7 + 2a, 12 - 2a, 2 + 2a, 2 + 2a)_{LR}$. Since, the fuzzy optimal value is depending upon '$a$' so for the chosen fully fuzzy linear programming problem infinite fuzzy numbers, representing the fuzzy optimal values, can be obtained.

Since, the fuzzy optimal values of the fully fuzzy linear programming problem, chosen in Example 6.1, corresponding to alternative fuzzy optimal solutions are not equal i.e., the results obtained by using the method, presented in Chap. 4, do not conform to the uniqueness property of fuzzy optimal value of a fully fuzzy linear programming problem. So, it is not genuine to apply the method, presented in Chap. 4, for solving fully fuzzy linear programming problems.

## 6.2   Kaur and Kumar's Method Based on RMDS Approach

Kumar et al. [3] used four parameters Rank, Mode, Divergence and Left spread for comparing $LR$ flat fuzzy numbers. It can be easily seen that if $\tilde{A}$ and $\tilde{B}$ are two $LR$ flat fuzzy numbers such that $\text{Rank}(\tilde{A}) = \text{Rank}(\tilde{B})$, $\text{Mode}(\tilde{A}) = \text{Mode}(\tilde{B})$, $\text{Divergence}(\tilde{A}) = \text{Divergence}(\tilde{B})$ and Left spread$(\tilde{A}) = $ Left spread$(\tilde{B})$ then $\tilde{A} = \tilde{B}$.

In this section, to overcome the limitations of the method, presented in Chap. 4, a method proposed by Kaur and Kumar [1, 2], based on RMDS approach [3], is presented for solving fully fuzzy linear programming problems.

### 6.2.1   RMDS Approach

In this section, the existing ranking approach [3] for comparing $LR$ flat fuzzy numbers is presented.

Let $\tilde{A} = (m_1, n_1, \alpha_1, \beta_1)_{LR}$ and $\tilde{B} = (m_2, n_2, \alpha_2, \beta_2)_{LR}$ be two $LR$ flat fuzzy numbers then use the following steps to compare $\tilde{A}$ and $\tilde{B}$:

**Step 1** Find Rank$(\tilde{A}) = \frac{1}{2}\left[\int_0^1 m_1 d\lambda - \int_0^1 \alpha_1 L^{-1}(\lambda)d\lambda + \int_0^1 n_1 d\lambda + \int_0^1 \beta_1 R^{-1}\right.$
$\left.(\lambda)d\lambda\right]$ and Rank$(\tilde{B}) = \frac{1}{2}\left[\int_0^1 m_2 d\lambda - \int_0^1 \alpha_2 L^{-1}(\lambda)d\lambda + \int_0^1 n_2 d\lambda + \right.$
$\left.\int_0^1 \beta_2 R^{-1}(\lambda)d\lambda\right]$

**Case (i)** If Rank$(\tilde{A}) >$ Rank$(\tilde{B})$ then $\tilde{A} \succ \tilde{B}$
**Case (ii)** If Rank$(\tilde{A}) <$ Rank$(\tilde{B})$ then $\tilde{A} \prec \tilde{B}$
**Case (iii)** If Rank$(\tilde{A}) =$ Rank$(\tilde{B})$ then go to Step 2.

**Step 2** Find Mode$(\tilde{A}) = \frac{1}{2}\int_0^1 (m_1 + n_1)d\lambda$ and Mode$(\tilde{B}) = \frac{1}{2}\int_0^1 (m_2 + n_2)d\lambda$
**Case (i)** If Mode$(\tilde{A}) >$ Mode$(\tilde{B})$ then $\tilde{A} \succ \tilde{B}$
**Case (ii)** If Mode$(\tilde{A}) <$ Mode$(\tilde{B})$ then $\tilde{A} \prec \tilde{B}$
**Case (iii)** If Mode$(\tilde{A}) =$ Mode$(\tilde{B})$ then go to Step 3

**Step 3** Find Divergence$(\tilde{A}) = \int_0^1 n_1 d\lambda + \int_0^1 \beta_1 R^{-1}(\lambda)d\lambda - \int_0^1 m_1 d\lambda + \int_0^1 \alpha_1 L^{-1}$
$(\lambda)d\lambda$ and Divergence$(\tilde{B}) = \int_0^1 n_2 d\lambda + \int_0^1 \beta_2 R^{-1}(\lambda)d\lambda - \int_0^1 m_2 d\lambda +$
$\int_0^1 \alpha_2 L^{-1}(\lambda)d\lambda$
**Case (i)** If Divergence$(\tilde{A}) >$ Divergence$(\tilde{B})$ then $\tilde{A} \succ \tilde{B}$
**Case (ii)** If Divergence$(\tilde{A}) <$ Divergence$(\tilde{B})$ then $\tilde{A} \prec \tilde{B}$
**Case (iii)** If Divergence$(\tilde{A}) =$ Divergence$(\tilde{B})$ then go to Step 4

**Step 4** Find Left spread$(\tilde{A}) = \int_0^1 \alpha_1 L^{-1}(\lambda)d\lambda$ and Left spread$(\tilde{B}) = \int_0^1 \alpha_2 L^{-1}$
$(\lambda)d\lambda$
**Case (i)** If Left spread$(\tilde{A}) >$ Left spread$(\tilde{B})$ then $\tilde{A} \succ \tilde{B}$
**Case (ii)** If Left spread$(\tilde{A}) <$ Left spread$(\tilde{B})$ then $\tilde{A} \prec \tilde{B}$
**Case (iii)** If Left spread$(\tilde{A}) =$ Left spread$(\tilde{B})$ then $\tilde{A} = \tilde{B}$.

### 6.2.2 Kaur and Kumar's Method

In this section, to overcome the limitations of the method, presented in Chap. 4, a method proposed by Kaur and Kumar [1, 2] is presented for solving fully fuzzy linear programming problems (4.1).

**Step 1** Use Step 1 to Step 8 of the method, presented in Chap. 4, and check that an alternative optimal solution of crisp linear programming problem (4.9) exist or not.

**Case (i)**   If an alternative optimal solution does not exist then the fuzzy optimal solution, obtained by using the method, presented in Chap. 4, is exact fuzzy optimal solution of the fully fuzzy linear programming problem (4.1).

**Case (ii)**   If alternative optimal solution exist then Go to Step 2.

**Step 2** Let the optimal value of the problem (4.9) be '$a$' and it occurs corresponding to '$p$' basic feasible solutions $\{x_j^k, y_j^k, \alpha_j''^k, \beta_j''^k\}$ where $k = 1, \ldots, p$. Now, our aim is to find $\max \text{ (or min)}\limits_{1 \le k \le p}\{\sum\limits_{j=1}^{n} (p_j, q_j, \alpha_j', \beta_j')_{LR} \otimes (x_j^k, y_j^k, \alpha_j''^k, \beta_j''^k)_{LR}\}$. Since,

$\text{Rank}(\sum\limits_{j=1}^{n} (p_j, q_j, \alpha_j', \beta_j')_{LR} \otimes (x_j^k, y_j^k, \alpha_j''^k, \beta_j''^k)_{LR}) = a \; \forall \; k = 1, \ldots, p$ so using

Step 2 of the RMDS approach, discussed in Sect. 6.2.1 if $\max \text{ (or min)}\limits_{1 \le t \le l}\{\text{Mode}(\sum\limits_{j=1}^{n} (p_j,$

$q_j, \alpha_j', \beta_j')_{LR} \otimes (x_j^k, y_j^k, \alpha_j''^k, \beta_j''^k)_{LR})\}$ is $\text{Mode}(\sum\limits_{j=1}^{n} (p_j, q_j, \alpha_j', \beta_j')_{LR} \otimes (x_j^\phi, y_j^\phi, \alpha_j''^\phi,$

$\beta_j''^\phi)_{LR})$ then $\max \text{ (or min)}\limits_{1 \le t \le l}\{\sum\limits_{j=1}^{n} (p_j, q_j, \alpha_j', \beta_j')_{LR} \otimes (x_j^k, y_j^k, \alpha_j''^k, \beta_j''^k)_{LR}\}$ will also

be $\sum\limits_{j=1}^{n} (p_j, q_j, \alpha_j', \beta_j')_{LR} \otimes (x_j^\phi, y_j^\phi, \alpha_j''^\phi, \beta_j''^\phi)_{LR}$ i.e., the fuzzy optimal solution of (4.1) can be obtained by solving the problem (6.1):

Maximize/Minimize $(\text{Mode}(\sum\limits_{j=1}^{n} (p_j, q_j, \alpha_j', \beta_j')_{LR} \otimes (x_j, y_j, \alpha_j'', \beta_j'')_{LR}))$

subject to

$$\sum_{j=1}^{n} m_{ij} = b_i \; \forall \; i = 1, 2, \ldots, m$$

$$\sum_{j=1}^{n} n_{ij} = g_i \; \forall \; i = 1, 2, \ldots, m$$

$$\sum_{j=1}^{n} \gamma_{ij}' = \gamma_i \; \forall \; i = 1, 2, \ldots, m \tag{6.1}$$

$$\sum_{j=1}^{n} \delta_{ij}' = \delta_i \; \forall \; i = 1, 2, \ldots, m$$

$$\text{Rank}(\sum_{j=1}^{n} \tilde{c}_j \otimes \tilde{x}_j) = a$$

$$x_j \le y_j, \alpha_j'' \ge 0, \beta_j'' \ge 0 \; \forall \; j = 1, 2, \ldots, n$$

**Case (i)**   If there does not exist any alternative optimal solution then the solution obtained from problem (6.1) is the optimal solution. Put the values of $x_j^*, y_j^*, \alpha_j''^*$ and $\beta_j''^*$ in $\tilde{x}_j^* = (x_j^*, y_j^*, \alpha_j''^*, \beta_j''^*)_{LR}$ to find the fuzzy optimal solution $\{\tilde{x}_j^*\}$ and find the fuzzy optimal value $\sum\limits_{j=1}^{n} (\tilde{c}_j \otimes \tilde{x}_j^*)$ by putting the values of $\tilde{x}_j^*$.

**Case (ii)**  If alternative solution exist then Go to Step 3.

**Step 3** On the same direction, as discussed in Step 2, solve the problem (6.2) and check that alternative optimal solution exist or not:

$$\text{Maximize/Minimize} \left( \text{Divergence} \left( \sum_{j=1}^{n} (p_j, q_j, \alpha'_j, \beta'_j)_{LR} \otimes (x_j, y_j, \alpha''_j, \beta''_j)_{LR} \right) \right)$$

subject to

$$\sum_{j=1}^{n} m_{ij} = b_i \ \forall i = 1, 2, \ldots, m$$

$$\sum_{j=1}^{n} n_{ij} = g_i \ \forall i = 1, 2, \ldots, m$$

$$\sum_{j=1}^{n} \gamma'_{ij} = \gamma_i \ \forall i = 1, 2, \ldots, m \tag{6.2}$$

$$\sum_{j=1}^{n} \delta'_{ij} = \delta_i \ \forall i = 1, 2, \ldots, m$$

$$\text{Rank} \left( \sum_{j=1}^{n} \tilde{c}_j \otimes \tilde{x}_j \right) = a$$

$$\text{Mode} \left( \sum_{j=1}^{n} \tilde{c}_j \otimes \tilde{x}_j \right) = b$$

$$x_j \leq y_j, \alpha''_j \geq 0, \beta''_j \geq 0 \ \forall j = 1, 2, \ldots, n$$

where '$b$' is the optimal value of the problem (6.1).

**Case (i)**  If there does not exist any alternative optimal solution then the solution obtained from problem (6.2) is the optimal solution. Put the values of $x_j^*, y_j^*, \alpha''^*_j$ and $\beta''^*_j$ in $\tilde{x}_j^* = (x_j^*, y_j^*, \alpha''^*_j, \beta''^*_j)_{LR}$ to find the fuzzy optimal solution $\{\tilde{x}_j^*\}$ and find the fuzzy optimal value $\sum_{j=1}^{n} (\tilde{c}_j \otimes \tilde{x}_j^*)$ by putting the values of $\tilde{x}_j^*$.

**Case (ii)**  If alternative solution exist then Go to Step 4.

**Step 4** On the same direction, as discussed in Step 2, solve the problem (6.3):

Maximize/Minimize (Left spread($\sum_{j=1}^{n}(p_j, q_j, \alpha'_j, \beta'_j)_{LR} \otimes (x_j, y_j, \alpha''_j, \beta''_j)_{LR})$)

subject to

$$\sum_{j=1}^{n} m_{ij} = b_i \ \forall i = 1, 2, \ldots, m$$

$$\sum_{j=1}^{n} n_{ij} = g_i \ \forall i = 1, 2, \ldots, m$$

$$\sum_{j=1}^{n} \gamma'_{ij} = \gamma_i \ \forall i = 1, 2, \ldots, m$$

$$\sum_{j=1}^{n} \delta'_{ij} = \delta_i \ \forall i = 1, 2, \ldots, m$$  (6.3)

$$\text{Rank}(\sum_{j=1}^{n} \tilde{c}_j \otimes \tilde{x}_j) = a$$

$$\text{Mode}(\sum_{j=1}^{n} \tilde{c}_j \otimes \tilde{x}_j) = b$$

$$\text{Divergence}(\sum_{j=1}^{n} \tilde{c}_j \otimes \tilde{x}_j) = c$$

$$x_j \leq y_j, \alpha''_j \geq 0, \beta''_j \geq 0 \ \forall j = 1, 2, \ldots, n$$

where '$c$' is the optimal value of the problem (6.2).

Find the fuzzy optimal solution $\{\tilde{x}^*_j\}$ by putting the values of $x^*_j, y^*_j, \alpha''^*_j$ and $\beta''^*_j$ in $\tilde{x}^*_j = (x^*_j, y^*_j, \alpha''^*_j, \beta''^*_j)_{LR}$ and fuzzy optimal value $\sum_{j=1}^{n}(\tilde{c}_j \otimes \tilde{x}^*_j)$ by putting the values of $\tilde{x}^*_j$.

## 6.3  Illustrative Example

In this section, to illustrate the method, presented in Sect. 6.2, fully fuzzy linear programming problem, chosen in Example 6.1, is solved. The exact fuzzy optimal solution of the chosen problem can be obtained by using the following steps:

**Step 1** Assuming $\tilde{x}_1 = (x_1, y_1, \alpha_1, \beta_1)_{LR}$, $\tilde{x}_2 = (x_2, y_2, \alpha_2, \beta_2)_{LR}$ and $\tilde{x}_3 = (x_3, y_3, \alpha_3, \beta_3)_{LR}$ and using Step 1 to Step 8 of the method, presented in Chap. 4, the fully fuzzy linear programming problem, chosen in Example 6.1, can be written as:

Maximize $\frac{1}{4}(2x_1 + 6y_1 + 4\beta_1 + 12x_2 + 26y_2 - 5\alpha_2 + 14\beta_2 + 12x_3 + 18y_3 - 5\alpha_3 + 10\beta_3)$

subject to

$$x_1 + x_2 = 1, y_1 + y_2 = 1, \alpha_1 + \alpha_2 = 0, \beta_1 + \beta_2 = 0$$
$$x_1 = x_3, y_1 = y_3, \alpha_1 = \alpha_3, \beta_1 = \beta_3$$
$$x_2 + x_3 = 1, y_2 + y_3 = 1, \alpha_2 + \alpha_3 = 0, \beta_2 + \beta_3 = 0$$
$$x_1 \geq 0, x_1 \leq y_1, \alpha_1 \geq 0, \beta_1 \geq 0$$
$$x_2 \geq 0, x_2 \leq y_2, \alpha_2 \geq 0, \beta_2 \geq 0$$
$$x_3 \geq 0, x_3 \leq y_3, \alpha_3 \geq 0, \beta_3 \geq 0$$

**Step 2** Since, on solving the crisp linear programming problem, obtained in Step 1, alternative optimal solution is obtained and the optimal value of the crisp linear programming problem is $\frac{19}{2}$ so, using Step 2 of the method, presented in Sect. 6.2, the solution of the chosen problem can be obtained by solving the following crisp linear programming problem:

$$\text{Maximize } (x_1 + \frac{7}{2}x_2 + \frac{7}{2}x_3 + y_1 + 6y_2 + 4y_3)$$

subject to

$$x_1 + x_2 = 1, y_1 + y_2 = 1, \alpha_1 + \alpha_2 = 0, \beta_1 + \beta_2 = 0$$
$$x_1 = x_3, y_1 = y_3, \alpha_1 = \alpha_3, \beta_1 = \beta_3$$
$$x_2 + x_3 = 1, y_2 + y_3 = 1, \alpha_2 + \alpha_3 = 0, \beta_2 + \beta_3 = 0$$
$$2x_1 + 6y_1 + 4\beta_1 + 12x_2 + 26y_2 - 5\alpha_2 + 14\beta_2$$
$$+ 12x_3 + 18y_3 - 5\alpha_3 + 10\beta_3 = 38$$
$$x_1 \geq 0, x_1 \leq y_1, \alpha_1 \geq 0, \beta_1 \geq 0$$
$$x_2 \geq 0, x_2 \leq y_2, \alpha_2 \geq 0, \beta_2 \geq 0$$
$$x_3 \geq 0, x_3 \leq y_3, \alpha_3 \geq 0, \beta_3 \geq 0$$

**Step 3** Since, on solving the crisp linear programming problem, obtained in Step 2, alternative optimal solution is obtained and the optimal value of the crisp linear programming problem is $\frac{19}{2}$ so, using Step 3 of the method, presented in Sect. 6.2, the solution of the chosen problem can be obtained by solving the following crisp linear programming problem:

Maximize $(4y_1 + 14y_2 + 10y_3 + 4\beta_1 + 14\beta_2 + 10\beta_3 - 5x_2 - 5x_3 + 5\alpha_2 + \alpha_3)$

subject to

$$x_1 + x_2 = 1, \; y_1 + y_2 = 1, \; \alpha_1 + \alpha_2 = 0, \; \beta_1 + \beta_2 = 0$$

$$x_1 = x_3, \; y_1 = y_3, \; \alpha_1 = \alpha_3, \; \beta_1 = \beta_3$$

$$x_2 + x_3 = 1, \; y_2 + y_3 = 1, \; \alpha_2 + \alpha_3 = 0, \; \beta_2 + \beta_3 = 0$$

$$2x_1 + 6y_1 + 4\beta_1 + 12x_2 + 26y_2 - 5\alpha_2 + 14\beta_2$$
$$+ 12x_3 + 18y_3 - 5\alpha_3 + 10\beta_3 = 38$$

$$x_1 + \frac{7}{2}x_2 + \frac{7}{2}x_3 + y_1 + 6y_2 + 4y_3 = \frac{19}{2}$$

$$x_1 \geq 0, \; x_1 \leq y_1, \; \alpha_1 \geq 0, \; \beta_1 \geq 0$$

$$x_2 \geq 0, \; x_2 \leq y_2, \; \alpha_2 \geq 0, \; \beta_2 \geq 0$$

$$x_3 \geq 0, \; x_3 \leq y_3, \; \alpha_3 \geq 0, \; \beta_3 \geq 0$$

**Step 4** Since, on solving the crisp linear programming problem, obtained in Step 3, alternative optimal solution is obtained and the optimal value of the crisp linear programming problem is 9 so, using Step 4 of the method, presented in Sect. 6.2, the solution of the chosen problem can be obtained by solving the following crisp linear programming problem:

Maximize $(2x_1 + 2x_2 + 2x_3 + 5\alpha_2 + 5\alpha_3)$

subject to

$$x_1 + x_2 = 1, \; y_1 + y_2 = 1, \; \alpha_1 + \alpha_2 = 0, \; \beta_1 + \beta_2 = 0$$

$$x_1 = x_3, \; y_1 = y_3, \; \alpha_1 = \alpha_3, \; \beta_1 = \beta_3$$

$$x_2 + x_3 = 1, \; y_2 + y_3 = 1, \; \alpha_2 + \alpha_3 = 0, \; \beta_2 + \beta_3 = 0$$

$$2x_1 + 6y_1 + 4\beta_1 + 12x_2 + 26y_2 - 5\alpha_2 + 14\beta_2$$
$$+ 12x_3 + 18y_3 - 5\alpha_3 + 10\beta_3 = 38$$

$$x_1 + \frac{7}{2}x_2 + \frac{7}{2}x_3 + y_1 + 6y_2 + 4y_3 = \frac{19}{2}$$

$$4y_1 + 14y_2 + 10y_3 + 4\beta_1 + 14\beta_2 + 10\beta_3 - 5x_2 - 5x_3 + 5\alpha_2 + \alpha_3 = 9$$

$$x_1 \geq 0, \; x_1 \leq y_1, \; \alpha_1 \geq 0, \; \beta_1 \geq 0$$

$$x_2 \geq 0, \; x_2 \leq y_2, \; \alpha_2 \geq 0, \; \beta_2 \geq 0$$

$$x_3 \geq 0, \; x_3 \leq y_3, \; \alpha_3 \geq 0, \; \beta_3 \geq 0$$

The obtained optimal solution is $x_1 = 1, \; y_1 = 1, \; \alpha_1 = 0, \; \beta_1 = 0, \; x_2 = 0, \; y_2 = 0, \; \alpha_2 = 0, \; \beta_2 = 0, \; x_3 = 1, \; y_3 = 1, \; \alpha_3 = 0$ and $\beta_3 = 0$. Using Step 4 of the method, presented in Sect. 6.2, the fuzzy optimal solution is $\tilde{x}_1 = (1, 1, 0, 0)_{LR}, \; \tilde{x}_2 = (0, 0, 0, 0)_{LR}, \; \tilde{x}_3 = (1, 1, 0, 0)_{LR}$ and the fuzzy optimal value is $(9, 10, 4, 4)_{LR}$.

**Table 6.1** Results of the chosen fully fuzzy linear programming problems

| Example | Fuzzy optimal value | |
|---|---|---|
| | Method presented in Chap. 4 | Method presented in this chapter |
| 4.1 | $(14, 48, 58, 50)_{LR}$ | $(14, 48, 58, 50)_{LR}$ |
| 4.2 | $(3, 8, 5, 10)_{LR}$ | $(3, 8, 5, 10)_{LR}$ |
| 6.1 | $(7 + 2a, 12 - 2a, 2 + 2a, 2 + 2a)_{LR}, 0 \leq a \leq 1$ | $(9, 10, 4, 4)_{LR}$ |

## 6.4 Advantages of the Kaur and Kumar's Method

As discussed in Sect. 6.1, by using the method, presented in Chap. 4, a unique fuzzy number, representing the fuzzy optimal value, is not obtained which do not conform to the uniqueness property of fuzzy optimal value of fully fuzzy linear programming problems. While, by using the method, presented in Sect. 6.2, always a unique fuzzy number, representing the fuzzy optimal value, will be obtained i.e., on solving the fully fuzzy linear programming problems by using the method, presented in Sect. 6.2, the uniqueness property of fuzzy optimal value will always be preserved.

## 6.5 Comparative Study

The results of the fully fuzzy linear programming problems, chosen in Examples 6.1, 4.1 and 4.2, obtained by using the method, presented in Chap. 4, and the method presented in this chapter, are shown in Table 6.1.

It is obvious from the results, shown in Table 6.1, that on solving the fully fuzzy linear programming problems, chosen in Examples 4.1 and 4.2, by using the method, presented in Chap. 4, a unique fuzzy number, representing the fuzzy optimal value, is obtained. While, on solving the fully fuzzy linear programming problem, chosen in Example 6.1, by using the method, presented in Chap. 4, infinite fuzzy numbers, representing the fuzzy optimal value of the same problem, are obtained which do not conform to the uniqueness property of fuzzy optimal value of a fully fuzzy linear programming problems. However, on solving all the chosen fully fuzzy linear programming problems by using the method presented in this chapter a unique fuzzy number, representing the fuzzy optimal, is obtained.

## 6.6 Conclusions

On the basis of present study, it can be concluded that it is better to use the method, presented in this chapter, as compared to the method, presented in Chap. 4, for solving fully fuzzy linear programming problems with equality constraints.

# References

1. Kaur, J., Kumar, A.: Unique fuzzy optimal value of fully fuzzy linear programming problems. Control Cybern. **41**, 171–182 (2012)
2. Kaur, J., Kumar, A.: A new method to find the unique fuzzy optimal value of fuzzy linear programming problems. J. Optim. Theory Appl. **156**, 529–534 (2013)
3. Kaur, P., Kumar, A.: A new ranking approach and its application for solving fuzzy critical path problems. S. Afr. J. Ind. Eng. (accepted)

# Chapter 7
# Future Scope

(i) The method presented in Chap. 6 can be used to find the unique fuzzy optimal value of fully fuzzy linear programming problems with equality constraints. However, this method cannot be used to find the unique fuzzy optimal value of fully fuzzy linear programming problems with inequality constraints. In future, it may be tried to develop a method for the same.

(ii) To find the fuzzy optimal solution of the fully fuzzy linear programming problems by using the presented methods, there is a need to convert the fully fuzzy linear programming problems into crisp linear programming problems. In future, it may be tried to develop a method which can be used directly to find the fuzzy optimal solution of the fully fuzzy linear programming problems without converting it into crisp linear programming problems.

© Springer International Publishing Switzerland 2016
J. Kaur and A. Kumar, *An Introduction to Fuzzy Linear
Programming Problems*, Studies in Fuzziness and Soft Computing 340,
DOI 10.1007/978-3-319-31274-3_7

Printed in the United States
By Bookmasters

Printed in the United States
By Bookmasters